Reactions in Condensed Phases: Kinetics and Thermodynamics

REACTIONS IN CONDENSED PHASES: KINETICS AND THERMODYNAMICS

G. E. ZAIKOV AND V. M. M. LOBO

EDITORS

Nova Science Publishers, Inc.

New York

Library of Congress Cataloging-in-Publication Data
Available Upon Request

ISBN: 1-59033-681-X.

400 Oser Ave, Suite 1600
Hauppauge, New York 11788-3619
Tele. 631-231-7269 Fax 631-231-8175
e-mail: Novascience@earthlink.net
Web Site: http://www.novapublishers.com

Printed in the United States of America

CONTENTS

Chapter 1

SOME ASPECTS OF THERMO-DEGRADATION OF PVC: CROSS LINKING OF MACROMOLECULES

V. M. Yanborisov*, K. S. Minsker*, G. E. Zaikov** and V. G. Zaikov**

*Bashkir State University, 32 Fruze st., Ufa, 450074 Russia

**N.M. Emanuel Institute of Biochemical Physics

4 Kosygin st., Moscow 119991 Russia

ABSTRACT

In the present work, the direct experimental proof is presented that cross-linking of macromolecules during thermo-degradation of PVC proceeds with participation of blocks of conjugated bonds formed at activation by carbonylallyl groups.

One of the reactions resulting in a break in conjugated polyene sequences formed during PVC,s degradation is the cross-linking reaction. It is assumed that in the absence of any active additives, which could activate the cross-linking reaction by a radical mechanism, this reaction proceeds according to the Diels-Alder mechanism (1-3). This conclusion emerges from the following experimental data:

- In the presence of active dienophiles (e.g. maleic anhydride, ethers of maleic acid, metal maleinates), capable of reacting with polyene blocks by a Diels-Alder reaction, the time of commencement of gel formation in degraded PVC increases (1);
- Upon treatment of a cross-linking polymer with maleic acid, a reduction of gels is observed due to diene "recondensation" of polyenes with the stronger dienophile (2); another conclusion emerges as well - cross-linking in PVC is a reversible reaction;
- Values of the kinetic parameters of the cross-linking reaction of PVC under thermo-degradation in vacuum – E_{act} = 50 ± 8 kJ/mol, lg A = 6.3 ± 0.5 (3) are close to the values known for the Diels-Alder reaction - E_{act} = 40 - 80 kJ/mol, lg A = 5 – 7 (4);

- Modification of PVC with conjugated dienes results in an increase of the time of the commencement of gel formation during thermo-degradation, in comparison to the unmodified polymer, owing to competition between the cross-linking reaction and the interaction of conjugated dienes with polyenes (5).

In (3, 6) it is assumed, that polyene sequences (ketopolyenes) with the general structure of ~C(O)-(-CH=CH-)$_n$~ take part in the cross-linking process, and each block of conjugated C=C bonds can act as both diene and dienophile according to the scheme:

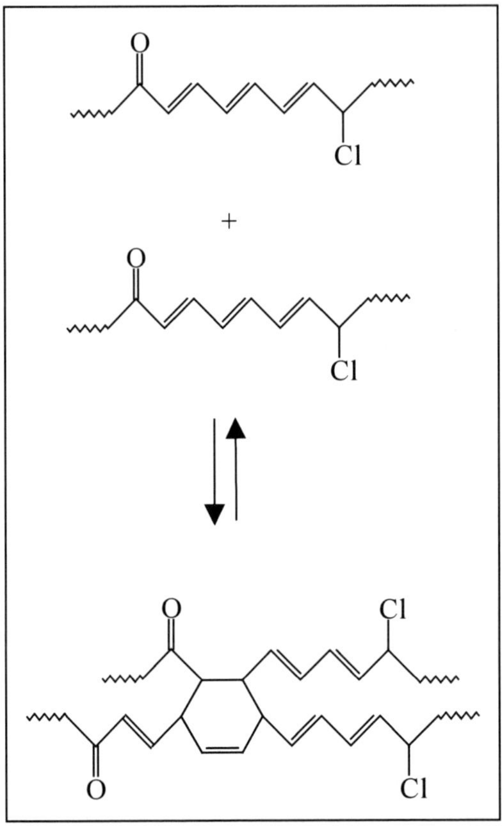

In the present work, the direct experimental proof that cross-linking of macromolecules during thermodegradation of PVC proceeds with the participation of blocks of conjugated bonds formed at activation by carbonylallyl groups is presented. If cross-linking of PVC macromolecules occurs with the participation of polyene sequences, whose formation is initiated by carbonylallyl groups, a depletion of sol-fraction by polyene sequences ~C(O)-(-CH=CH-)$_n$~ should occur upon accumulation of a gel-fraction. The amount of ketopolyenes γ_0 is easily determined by a method of hydrolytic cleavage on change of intrinsic viscosity of the sol-fraction after alkaline hydrolysis of PVC and subsequent calculation with hydrolysis of PVC and subsequent calculation with the formula (7):

$$\gamma_0 = \frac{62.5 \times 1.86}{M_\eta^0} \times \left\{ \left(\frac{\eta_0}{\eta} \right)^{1.38} - 1 \right\} \qquad (1)$$

where η_0 and η - intrinsic viscosity before and after the macromolecule's cleavage, M_η^0 - initial viscosity-average molecular weight.

Indeed, the experimental data (Fig. 1) reveal that gradual decrease of ketopolyene contents in the sol-fraction is observed with an increase of duration of the degradation of PVC. Hence we can postulate that the polyene sequence, those formation is initiated by carbonylallyl groups, undoubtfully participates in the cross-linking reaction of the macromolecules during degradation of PVC.

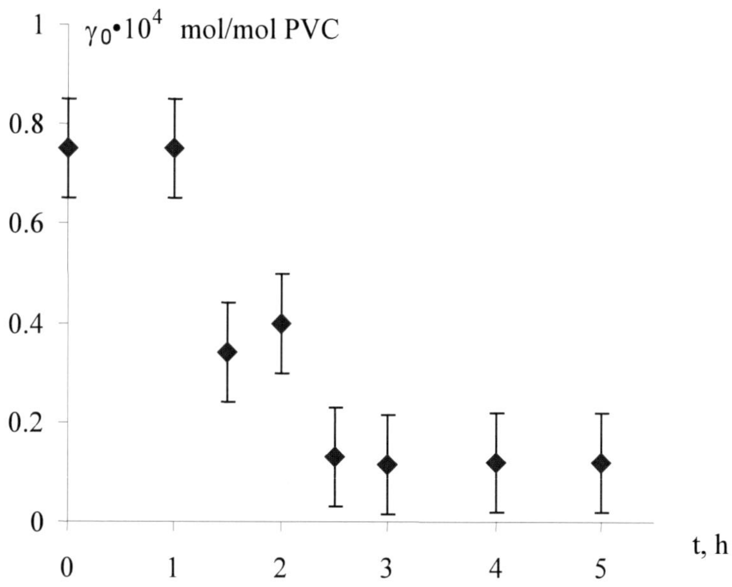

Fig 1. Dependence of carbonylallyl groups contents in sol-fraction of degraded PVC (PVC-C-70, $\overline{M}_\eta = 110000$) of degradation duration.

However, the cross-linking reaction of the ketopolyene sequences along can not ensure an experimentally observable yield of a gel under prolonged degradation which considerably exceeds the time of the beginning of gel formation of (τ) (6, 8), since the fraction of PVC macromolecules containing is small (8-10 mol %). Therefore, at prolonged stages of PVC degradation, the reaction of diene recondensation with polyene blocks can occur. Those polyene blocks are formed as a result of the activation of the cleavage of macromolecules by β-chloroallyl groups, which were formed via the elimination of HCl from regular vinyl chloride units and *tert*-chloride groups (3). In PVC degradation studies in the presence of barium stearate as HCl acceptor, the secondary reaction of possible readdition of HCl to the polyene sequences is suppressed (9). The most probable secondary reactions under these conditions are, obviously, the reactions of macromolecular cross-linking (1-5) and cyclization of polyenes (10). In (6) it was found that the time of commencement of gel formation depends on the content of carbonylallyl groups (Fig 2, 1), while the PVC degradation studies in the work were carried out undo conditions of the removal of HCl from the reaction zone by freezing.

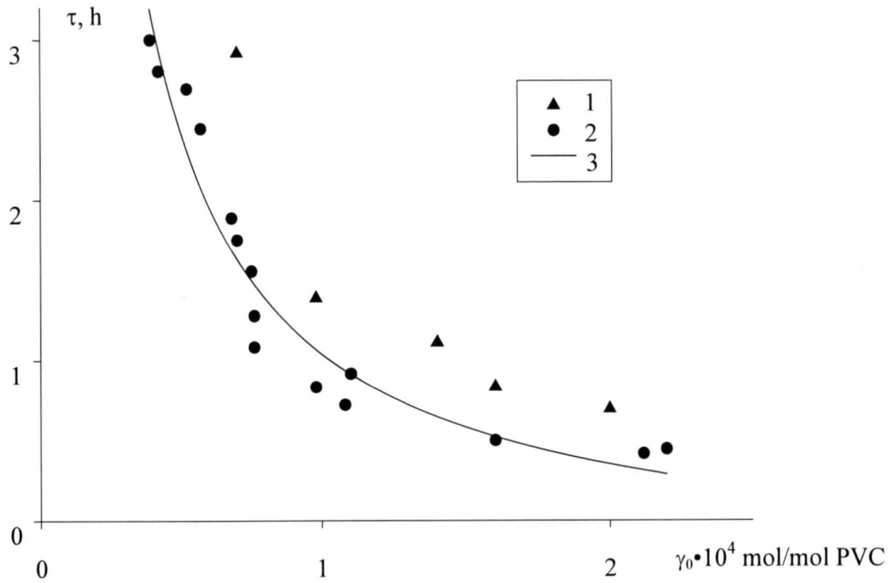

Fig 2. Dependence of commencement of gel formation on carbonylallyl groups contents in PVC: 1 – 448 K, 10^{-2} Pa, continuous removal of HCl (6), 2 – 448 K, 10^{-2} Pa, barium stearate; 3 – according to formula (3) at n = 1.26.

In order to develop the equation in the best way to describe the dependence $\tau = f(\gamma_0)$, it is necessary to investigate experimentally this dependence in a wide interval of γ_0. Meanwhile, the values of γ_0 for regular industrial and laboratory PVC samples are within very narrow limits – $0.8 \div 2.4 \times 10^{-4}$ mol/mol PVC. Therefore, the samples with small contents of carbonylallyl groups were specially obtained for this research by chemical modification of PVC with maleic anhydride (MA) (11). The reaction results in irreversible reduction of the content of carbonylallyl groups in the PVC and, apparently, to a delay of its dehydrochlorination rate as a result of reduction of the number of formed polyene sequences. The characteristics of the modified polymer are given in Table 1. At low levels of carbonylallyl groups in the polymer, the experimental error in viscometric determination of γ_0 according to the equation (1) is rather high, therefore in some cases values of γ_0^{mod} of modified PVC samples were calculated more precisely by the formula

$$\gamma_0 = (V_{HCl} - V_c)/k_p \qquad (2)$$

Rate of gross-hydrochlorination V_{HCl} was determined experimentally (Table 1), $V_c = 0.8 \cdot 10^{-7}$ mol/mol PVC•s^{-1} is rate of HCl elimination from regular PVC units, $k_p = 0.75 \cdot 10^{-2}$ s^{-1} is a constant of HCl elimination from chloropolyene sequences (3).

Table 1. Characteristics of PVC Samples, Modified with Maleic Anhydride

$\overline{M}_\eta^0$	$\gamma_0^{init} \cdot 10^4$ mol/mol PVC	[MA]•10^2, mol/mol PVC	T_{mod}, K	Modification duration, s	$[\eta]_0$, dL/g	$[\eta]$, dL/g	$\gamma_0^{mod} \cdot 10^4$ mol/mol PVC	V_{HCl}•10^3, mol/mol PVC (448 K)	τ, c (448 K)
110000	0.76	1	373	2400	1.09	1.04	0.70	0.61	6300
110000	0.76	1	373	3600	1.09	-	0.68*	0.59	6800
105000	0.80	1	383	4800	1.07	-	0.75*	0.64	5600
110000	0.76	1	373	12000	1.09	1.05	0.57	0.51	8800
110000	0.76	4	373	10800	1.09	-	0.52*	0.47	9700
110000	0.76	4	398	7200	1.09	-	0.42*	0.40	10100

* Values of γ_0^{mod} were calculated by formula (3)

The typical dependence of the amount of gel formed as a function of the duration of degradation of PVC has a linear pattern (Fig. 3). The time of commencement of gel formation was determined by linear extrapolation at the intersection with an abscissa axis.

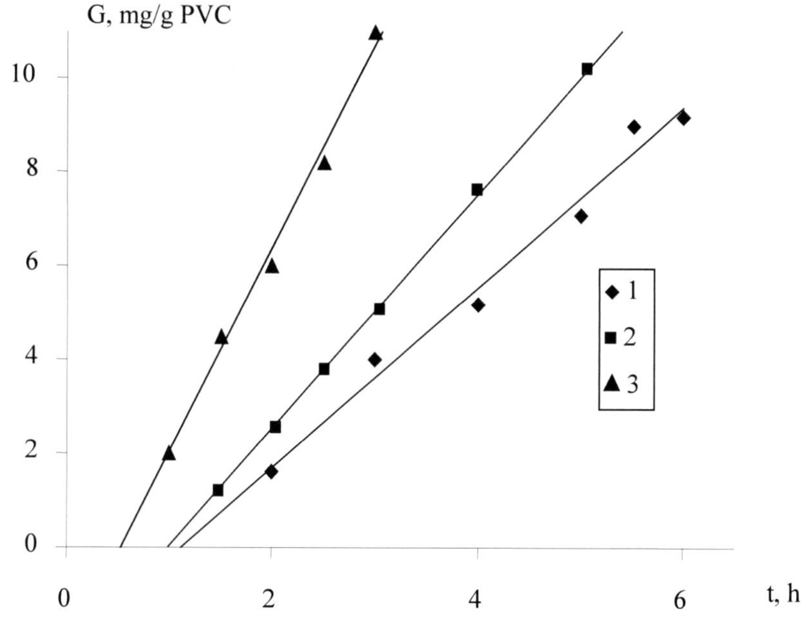

Fig 3. Accumulation of gel a PVC degradation (10^{-2} Pa, barium stearate). Carbonylallyl groups contents in the sample: 1 – 0.76•10^{-4} mol/mol PVC, 2 – 0.98•10^{-4} mol/mol PVC, 3 – 1.6•10^{-4} mol/mol PVC.

The presence of barium stearate in a PVC degradation study significantly inhibits the readdition of HCl to polyenes. This conclusion comes from the observed reduction of the of commencement time in comparison with the previous case (Fig. 2. 2).

In a general view, the dependence $\tau = f(\gamma_0)$ can be represented as

$$\tau = C \bullet \gamma_0^{-n} \tag{3}$$

Here n is the order of the macromolecular cross-linking reaction of blocks of conjugated bonds. Other values were obtained by the method of least squares, with $n=1.26 \pm 0.17$ at correlation coefficient 0.979 with confidential probability 95%.

Thus, the order of the reaction on ketopolyenes is close to one, therefore it is more probable that macromolecular cross-linking during PVC thermodegradation is caused, not by the interaction of polyene blocks among themselves, as it was assumed earlier (6), but mainly by the interaction of polyenes with other unsaturated groups in PVC. Such groups can be unsaturated end-groups, chose contents in PVC are about 10^{-3} mol/mol PVC (2) and internal β-chloroallyl groups formed during PVC's thermodegradation at statistical elimination of HCl from regular units and *tert*-chloride groups.

Modification of PVC with maleic anhydride was carried out as described in (11). PVC degradation was studied at 448 K in sealed ampoules, at a residual pressure of 10^{-2} Pa. Liberckod HCl was absorbed by barium stearate (stabilizer – acceptor of HCl) and was quantitatively determined by mercurimetry. Gel-fraction in samples of degraded PVC was separated by subsequent washing and centrifuging of the polymer with fresh cyclohehanone in the same ampoules where the degradation of PVC took place. In order to remove unreacted barium stearate, the precipitate was washed three times with dodecyl alcohol at 373 K. The isolated gel was washed three times with diethyl ether in order to remove the residual solvent and who subsequent to constant weight at 348 K. Experimental error in the quantitative determination by the gel did not exceed 10 %. The time of commencement of gel formation was determined by linear extrapolation of the dependence G = f(t) to the abscissa axis. Alkaline hydrolysis of PVC samples and measurement of its intrinsic viscosity in order to determine the amount of carbonylallyl groups was done according to (7). Sol-fraction of degraded PVC was obtained by centrifugation during 1 h at 5000 RPM.

References

[1] Razuvaev, G. A.; Troitskaya, L. S.; Troitskii, B. B. *J. Polym. Sci. A-1* **9(9)**, 2673 (1971).

[2] Kelen, T.; Iván, B.; Nagy, T,; Nurcsanyi, B.; Kennedy, J. P. *Polym. Bull.* **1(2)**, 79 (1978).

[3] Minsker, K. S.; Zaikov, G. E. *Chemistry of Chlorine – Containing Polymers: Synthesis, Degradation, Stabilization*; NOVA Science Publishers: New York (2000).

[4] Ingold, K. *Theoretical bases of organic chemistry*; Mir: Moscow (1973).

[5] Minsker, K. S.; Kolesov, S. V.; Petrov, V. V. *Vysokomol. Soedin.* **25A(4)**, 732 (1983).

[6] Berlin, Al. Al.; Minsker, K. S.; Kolesov, S. V.; Balandina, N. A. *Vysokomol. Soedin.* **19B(2)**, 132 (1977).

[7] Lisitskii, V. V.; Kolesov, S. V.; Gataullin, R. F.; Minsker, K. S. J. *Analit. Chem.* **33(11)**, 2202 (1978).

[8] Kelen, T. J. *Macromol. Sci.* **12(3)**, 349 (1978).

[9] Oleinik, E. P.; Vasileiskaya, N. S.; Razuvaev, G. A. Izv. *Akad. Nauk SSSR, Ser. Khim.* **3**, 482 (1968).

[10] Troitskii, B. B.; Troitskaya, L. S.Int. *J. Polym. Mater.* **41**, 285 (1998).

[11] Minsker, K. S.; Kolesov, S. V.; Petrov, V. V. *Dokl. Akad. Nauk SSSR* **268(3)**, 632 (1983).

Chapter 2

QUANTUM CHEMICAL CALCULATION OF 4-M ETHYL -2, 6-DI-TRET-BUTHYLPHENOL (IONOLE)

V. A. Babkin, S. K. Minsker, Yu. A. Sangalov,
R. G. Fedunov, and A. I. Rahimov
Bushkirian State University, 32 Franze St., UFA-450074, Russia;
G. E. Zaikov
N.M.Emanuel Institute of Biochemical Physics, 4, Kosygin St.,
Moscow 117-334, Russia.

ABSTRACT

The calculation of molecule of stabilizer-antioxidant of 4-methyl-2,6-di-tret-butylphenol (ionole) has been carried out by the quantum-chemical semi-empirical MNDO method in Dewar and Teel parameterization. The geometrical and electronic structure of this combination was obtained. The ionole acidity strength has been estimated by theoretical using the express-method of estimating acid strength of H-acids and has been measured by experimental. We found, that 4-methyl-2,6-di-tret-butylphenol is to rather weak H-acids (pKa=14.15, pH=6.4-6.8, where pKa is universal acidity index, pH=-lg[H^+].

AIMS AND BACKGROUND

The electronic and geometrical structure of ionole, the acidity strength are most important characteristics defining their physical-chemical characteristics in different chemical reaction, for example in polymer-analogous transformations of polyolefins and polydiens with phenols in presence of electrophilic catalysts of HEMeAICUOH] type (where Me=Li,Na,K), which are effective catalysts of selective modification of polymeric hydrocarbons by phenols. They evince high activity and selectivity of action in process of alkyling of phenols (reaction pass on classical mechanism of electrophylic substitution) and other electrophylic reactions which are rigidly connected with acidity strength of components of catalytic system [1-3].

And also important is comprehension that ionole is one of the best antioxidant additives [4] of synthetic rubber, polyolefins, PVC (polyvynilchloride), rocket fuels [5].

The aim of work is quantum-chemical calculation of molecule of ionole by quantum-chemical MNDO method, theoretical and experimental estimation his acidity strength.

METHODOLOGY

To calculate the model we used MNDO method in Dewar and Teel parameterization. The minimization of total energy of the system has been performed at all geometric parameters by Davidon-Flatcher-Powell method [7]. This method gives more correctly relative stability of structures and values of valent angles than other methods. Initial values of bond lengths and angles have been taken from reference. The calculation of ionole has been performed in the ground state in classical approximation of isolated molecule in a gas phase which always has been considered enough correct for unpolar indifferent solvents. Therefore $Eq_{sys}=0$ (where q_{sys} is total charge of molecular system), multiplicity is M=2S+1 (where S is summary spin of molecules in the ground state equal to 0).

To estimate theoretically acid strength of ionole we used express-method offered in [6], that consists in quantum-chemical calculation of H-acid and definition of the maximum charge on hydrogen atom $q_{H+}{}^{max}$ and calculation of pK_a by formula $pK_a=42.11-147.18\ q_{H+}{}^{max}$.

The acid strength of ionole has been defined experimentally **in** different solvent by pH-meter OP-265/1 combined by glass electrode OP-0808 (maker is Radelkis Budapest).

RESULTS AND DISCUSSION

The optimized geometric and electronic structure of 4-methyl-2,6-di-tret-buthylphenol (ionole) is shown in Fig. 1. The values of optimized geometric parameters (bond lengths, valent and torsion angles) is represented in Table 1. It is noted the rather high values of E_o=-59504.92 kcal/mole and E_B=-3901.89 kcal/mole (E_B is summary energy of ionole bond). The hydrogen atom N40 (see Fig. 1) possesses the maximum positive charge $q_{H+}{}^{max} = +0.19$. The value of universal acidity index of ionole pKa is possible to estimate by formula:

$$pKa=42.11-147.18q_{H+}{}^{max}=42.11 -147.10*0.19= 14.15$$

pKa(ionole) is 14.15.

The acidity strength of ionole measured experimentally in different solvents correspondingly is equal to: pH(H20) = 6.40; pH(C2H50H) = 6.76; pH(toleole) = 6.79; pH(dioxan) = 6.80; pH(CCl4) = 6.73.

Thus quantum-chemical calculation of 4-methyl-2,6-di-tret-buthylphenol (ionole) by semi-empiric MNDO method in Dewar and Teel parameterization with optimization of geometry of all parameters by Davidon-Flatcher-Pawell method allowed us to find their optimal geometric and electronic structure, and to estimate the acidity strength.

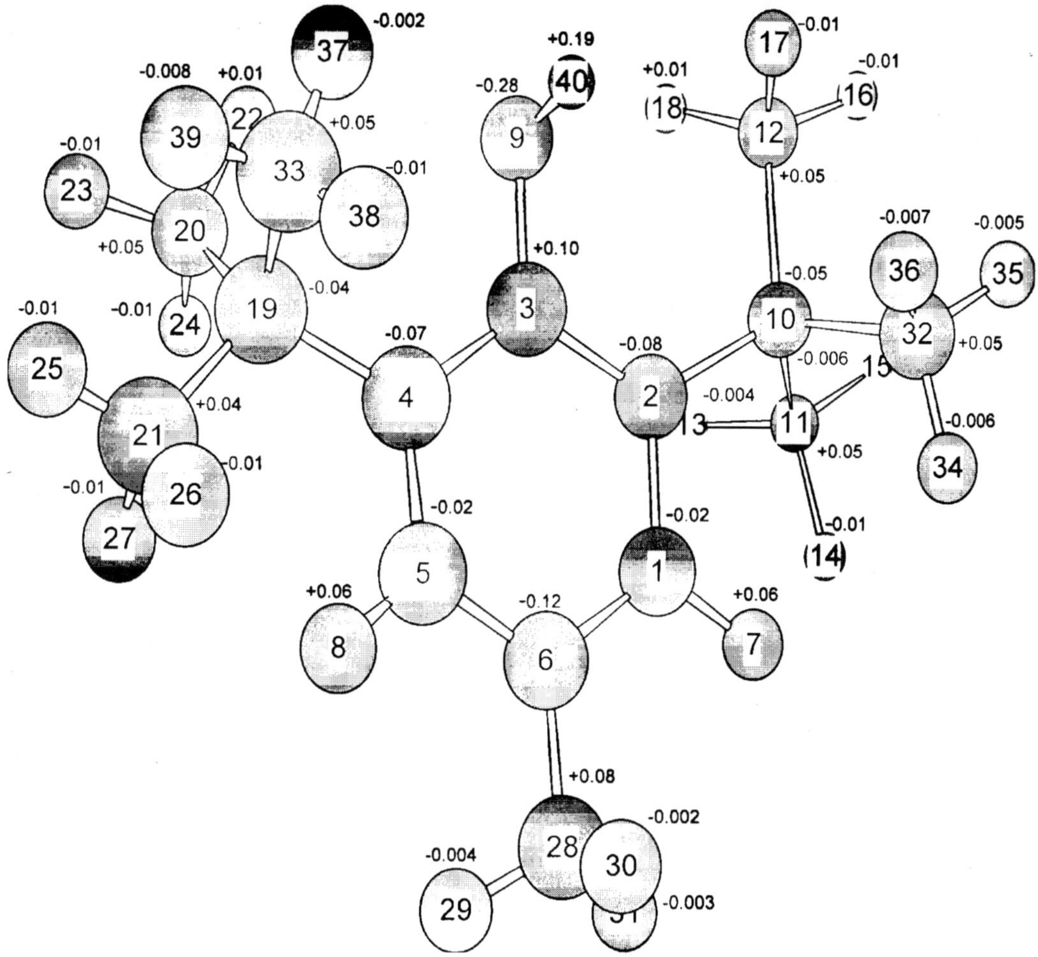

Fig. 1. The optimized geometric and electronic structure of molecule of 4-methyl-2,6-di-tret-bolecule of4-methyl-2,6-di-tret-be

Table 1. The Optimized Values of Bond Lengths, Valent and Torsion Angles of Ionole Molecule

Bond length, A	Valent angle, degree	Torsion angle, degree
C(1)-C(2) 1.420	C(2)-C(1)-C(6) 123.297	C(6)-C(1)-C(2)-C(3) 3.394
C(1)-C(6) 1.404	C(2)-C(1)-H(7) 119.984	C(6)-C(1)-C(2)-C(10) -176.517
C(1)-H(7) 1.093	C(6)-C(1)-H(7) 116.720	H(7)-C(1)-C(2)-C(3) -176.378
C(2)-C(3) 1.434	C(1)-C(2)-C(3) 115.873	H(7)-C(1)-C(2)-C(10) 3.691
C(2)-C(10) 1.553	C(1)-C(2)-C(10) 116.296	C(2)-C(1)-C(6)-C(5) 1.343
C(3)-C(4) 1.444	C(3)-C(2)-C(10) 127.831	C(2)-C(1)-C(6)-C(28) -178.296
C(3)-0(9) 1.367	C(2)-C(3)-C(4) 122.967	H(7)-C(1)-C(6)-C(5) -178.902
C(4)-C(5) 1.410	C(2)-C(3)-0(9) 119.596	H(7)-C(1)-C(6)-C(28) 1.518
C(4)-C(19) 1.553	C(4)-C(3)-0(9) 117.391	C(1)-C(2)-C(3)-C(4) -7.751
C(5)-C(6) 1.410	C(3)-C(4)-C(5) 116.132	C(1)-C(2)-C(3)-0(9) 174.769
C(5)-H(8) 1.091	C(3)-C(4)-C(19) 124.034	C(10)-C(2)-C(3)-C(4) 172.158

Bond length, A	Valent angle, degree	Torsion angle, degree
C(6)-C(28) 1.507	C(5)-C(4)-C(19)119.827	C(10)-C(2)-C(3)-0(9) -5.317
0(9)-H(40) 0.942	C(4)-C(5)-C(6) 123.100	C(1)-C(2)-C(10)-C(11) 55.107
C(10)-C(11) 1.564	C(4)-C(5)-H(8) 120.907	C(1)-C(2)-C(10)-C(12) 173.909
C(10)-C(12) 1.559	C(6)-C(5)-H(8) 115.992	C(1)-C(2)-C(10)-C(32) -65.115
C(10)-C(32) 1.561	C(1)-C(6)-C(5) 118.192	C(3)-C(2)-C(10)-C(11) -124.803
C(11)-H(13) 1.109	C(1)-C(6)-C(28) 120.864	C(3)-C(2)-C(10)-C(12) -6.001
C(11)-H(14) 1.108	C(5)-C(6)-C(28) 120.944	C(3)-C(2)-C(10)-C(32) 114.974
C(11)-H(15) 1.110	C(3)-0(9)-H(40) 110.846	C(2)-C(3)-C(4)-C(5) 7.076
C(12)-H(16) 1.112	C(2)-C(10)-C(11) 108.528	C(2)-C(3)-C(4)-C(19) -171.880
C(12)-H(17) 1.108	C(2)-C(10)-C(12) 117.195	0(9)-C(3)-C(4)-C(5) -175.400
C(12)-H(18) 1.107	C(2)-C(10)-C(32) 108.529	0(9)-C(3)-C(4)-C(19) 5.646
C(19)-C(20) 1.561	C(11)-C(10)-C(12) 105.132	C(2)-C(3)-0(9)-H(40) -80.908
C(19)-C(21) 1.564	C(11)-C(10)-C(32) 110.580	C(4)-C(3)-0(9)-H(40) 101.476
C(19)-C(33) 1.559	C(12)-C(10)-C(32) 106.786	C(3)-C(4)-C(5)-C(6) -1.899
C(20)-H(22) 1.106	C(10)-C(11)-H(13) 111.012	C(3)-C(4)-C(5)-H(8) 177.696
C(20)-H(23) 1.111	C(10)-C(11)-H(14) 113.231	C(19)-C(4)-C(5)-C(6) 177.082
C(20)-H(24) 1.109	C(10)-C(11)-H(15) 111.516	C(19)-C(4)-C(5)-H(8) -3.320
C(21)-H(25) 1.111	H(13)-C(11)-H(14) 107.433	C(3)-C(4)-C(19)-C(20) 60.468
C(21)-H(26) 1.109	H(13)-C(11)-H(15) 107.452	C(3)-C(4)-C(19)-C(21) 178.450
C(21)-H(27) 1.109	H(14)-C(11)-H(15) 105.866	C(3)-C(4)-C(19)-C(33) -62.774
C(28)-H(29) 1.109	C(10)-C(12)-H(16) 110.650	C(5)-C(4)-C(19)-C(20) -118.447
C(28)-H(30) 1.111	C(10)-C(12)-H(17) 112.675	C(5)-C(4)-C(19)-C(21) -0.448
C(28)-H(31) 1.109	C(10)-C(12)-H(18) 112.486	C(5)-C(4)-C(19)-C(33) 118.311
C(32)-H(34) 1.108	H(16)-C(12)-H(17) 105.337	C(4)-C(5)-C(6)-C(1) -2.100
C(32)-H(35) 1.110	H(16)-C(12)-H(18) 106.337	C(4)-C(5)-C(6)-C(28) 177.528
C(32)-H(36) 1.109	H(17)-C(12)-H(18) 108.923	H(8)-C(5)-C(6)-C(1) 178.296
C(33)-H(37) 1.107	C(4)-C(19)-C(20) 109.722	H(8)-C(5)-C(6)-C(28) -2.076
C(33)-H(38) 1.109	C(4)-C(19)-C(21) 113.016	C(1)-C(6)-C(28)-H(29) -153.808
C(33)-H(39) 1.110	C(4)-C(19)-C(33) 110.977	C(1)-C(6)-C(28)-H(30) 86.208
	C(20)-C(19)-C(21) 105.943	C(1)-C(6)-C(28)-H(31) -33.743
	C(20)-C(19)-C(33) 111.161	C(5)-C(6)-C(28)-H(29) 26.576
	C(21)-C(19)-C(33) 105.880	C(5)-C(6)-C(28)-H(30) -93.406
	C(19)-C(20)-H(22) 113.530	C(5)-C(6)-C(28)-H(31) 146.640
	C(19)-C(20)-H(23) 111.210	C(2)-C(10)-C(11)-H(13) 55.420
	C(19)-C(20)-H(24) 110.988	C(2)-C(10)-C(11)-H(14) -65.519
	H(22)-C(20)-H(23) 106.012	C(2)-C(10)-C(11)-H(15) 175.210
	H(22)-C(20)-H(24) 107.491	C(12)-C(10)-C(11)-H(13) -70.735
	H(23)-C(20)-H(24) 107.274	C(12)-C(10)-C(11)-H(14) 168.324
	C(19)-C(21)-H(25) 111.119	C(12)-C(10)-C(11)-H(15) 49.050
	C(19)-C(21)-H(26) 112.282	C(32)-C(10)-C(11)-H(13) 174.361
	C(19)-C(21)-H(27) 112.357	C(32)-C(10)-C(11)-H(14) 53.419
	H(25)-C(21)-H(26) 106.253	C(32)-C(10)-C(11)-H(15) -65.857
	H(25)-C(21)-H(27) 106.195	C(2)-C(10)-C(12)-H(16) -173.909
	H(26)-C(21)-H(27) 108.262	C(2)-C(10)-C(12)-H(17) 68.458

Bond length, A	Valent angle, degree	Torsion angle, degree
	C(6)-C(28)-H(29) 112.149	C(2)-C(10)-C(12)-H(18) -55.136
	C(6)-C(28)-H(30) 110.362	C(11)-C(10)-C(12)-H(16) -53.309
	C(6)-C(28)-H(31) 111.923	C(11)-C(10)-C(12)-H(17) -170.938
	H(29)-C(28)-H(30) 107.625	C(11)-C(10)-C(12)-H(18) 65.467
	H(29)-C(28)-H(31) 106.839	C(32)-C(10)-C(12)-H(16) 64.206
	H(30)-C(28)-H(31) 107.718	C(32)-C(10)-C(12)-H(17) -53.426
	C(10)-C(32)-H(34) 112.812	C(32)-C(10)-C(12)-H(18) -177.014
	C(10)-C(32)-H(35) 111,563	C(2)-C(10)-C(32)-H(34) 64.129
	C(10)-C(32)-H(36) 111.138	C(2)-C(10)-C(32)-H(35) -176.187
	H(34)-C(32)-H(35) 106.379	C(2)-C(10)-C(32)-H(36) -56,327
	H(34)-C(32)-H(36) 107.216	C(11)-C(10)-C(32)-H(34) -54.808
	H(35)-C(32)-H(36) 107.421	C(11)-C(10)-C(32)-H(35) 64.872
	C(19)-C(33)-H(37) 113.424	C(11)-C(10)-C(32)-H(36) -175.262
	C(19)-C(33)-H(38) 111.254	C(12)-C(10)-C(32)-H(34) -168.672
	C(19)-C(33)-H(39) 111.521	C(12)-C(10)-C(32)-H(35) -48.990
	H(37)-C(33)-H(38) 107.708	C(12)-C(10)-C(32)-H(36) 70.873
	H(37)-C(33)-H(39) 105.459	C(4)-C(19)-C(20)-H(22) -71.837
	H(38)-C(33)-H(39) 107.092	C(4)-C(19)-C(20)-H(23) 168.715
		C(4)-C(19)-C(20)-H(24) 49.371
		C(21)-C(19)-C(20)-H(22) 165.862
		C(21)-C(19)-C(20)-H(23) 46.411
		C(21)-C(19)-C(20)-H(24) -72.931
		C(33)-C(19)-C(20)-H(22) 51.298
		C(33)-C(19)-C(20)-H(23) -68.152
		C(33)-C(19)-C(20)-H(24) 172.505
		C(4)-C(19)-C(21)-H(25) -178.902
		C(4)-C(19)-C(21)-H(26) 62.274
		C(4)-C(19)-C(21)-H(27) -60.050
		C(20)-C(19)-C(21)-H(25) -58.702
		C(20)-C(19)-C(21)-H(26) -177.547
		C(20)-C(19)-C(21)-H(27) 60.125
		C(33)-C(19)-C(21)-H(25) 59.438
		C(33)-C(19)-C(21)-H(2б) -59.411
		C(33)-C(19)-C(21)-H(27) 178.266
		C(4)-C(19)-C(33)-H(37) 79.257
		C(4)-C(19)-C(33)-H(38) -42.358
		C(4)-C(19)-C(33)-H(39) -161.853
		C(20)-C(19)-C(33)-H(37) -43.152
		C(20)-C(19)-C(33)-H(38) -164.769
		C(20)-C(19)-C(33)-H(39) 75.738
		C(21)-C(19)-C(33)-H(37) -157.756
		C(21)-C(19)-C(33)-H(38) 80.628
		C(21)-C(19)-C(33)-H(39) -38.868

It was shown, that ionole is rather weak H-acid, and this conforms to high selectivity of polymer-analogous transformation of polyolefins and polydiens with various phenols. The electron density distribution on atoms, the energy parameters (Eo, Eв) and other data of quantum-chemical calculation are able to help more profound comprehension of fact that ionole is standard of antioxidant additives. As proof of this aspect, it is necessary to note, that ionole possesses the maximum negative charge on oxigen atom q_{0-}^{max} =-0.28 in sequence of estimated phenols (4-iso-propyl-phenol, 2-benzylidenaminophenol, 4-benzylidenaminophenol. 2-methyl-phenol, 3-methyl-phenol, o-dioxabenzene, m-dioxabenzen, n-dioxabenzene and other) and given fact is necessary to consider to explain the unique characteristics of ionole.The Authors express thanks doktorant of Bashgosuniversitet (Ufa) Mufteev A.F. for performing the experimental part of work.

REFERENCES

[1] Babkin V.A., Zaikov G.E., Minsker K.S. *Quantum-chemical aspects of the cationic polymerization of the olefins*. 1997, New-York, 138 p.

[2] Babkin V.A., Zaikov G.E., Minsker K.S. *Kvantovo-khimicheskie aspekhty khationnoi polimerizatzii olefinov*. Ufa.: Filem, p. 187.

[3] Minsker K.S., Ivanova S.R., Babkin V.A. *0 sviazi aktivnosti i selectivnosti elektrofilnyh processov s meroi ih kislotnoi sily*. - DAN, 1995, v.343. N5, p.646-650.

[4] Kuliev A.M. *Khimia i tehnologia prisadok k maslam i toplivam*. - M, Khimia, 1972-160 p.

[5] Khimicheskaia enciclopedia. - M, 1988, v.1, p. 179.

[6] Babkin V.A., Sangalov Yu.A., Minsker K.S., Ponomarev O.A.. Fedunov R.G., Malyshev A.G., Zaikov G.E. Teoreticheskaia ocenka kislotnoi sily kompleksov "ftoristyi vodorod - ftorid boora" // *Bush.khim. jornal*. -1998, - v.5, - N5, p.52-53.

[7] Klark T. *Komputernaia khimia*. -1991. - p. 185.

Chapter 3

THE STRUCTURAL TREATMENT OF THE A NUMBER OF EFFECTIVE CENTERS OF POLYMERIC CHAINS IN THE PROCESS OF THE THERMOOXIDATIVE DEGRADATION

G. V. Kozlov and H. M. Batyrova
Kabardino-Balkarian State University
Chernishevsky st. 173, 360004, Nalchik, Russia
G. E. Zaikov
Institute of Biochemical Physics Russian Academy of Sciences
Kosygin st. 4, 119991, Moscow, Russia

ABSTRACT

It is shown that the effective number of the polymeric macromolecule reactive centers depends on the macromolecular coil structure. The quantitative characteristic of the latter one can be received within the framework of the fractal analysis. The increase of macromolecular coil fractal dimension meaning the rise of a degree of its compactness reduces in lowering the effective number of the reactive (accessible to the molecules of oxidant) sites of a polymeric macromolecule.

Keywords: Thermooxidative degradation, copolymers, macromolecular coil, fractal analysis, scaling approach, active centers.

INTRODUCTION

Now it is supposed that the number of the reactive centers of a polymeric macromolecule which are capable to react with the oxidant molecules, is defined by the chemical constitution of polymer [1]. However, the polymeric macromolecule both in a solution and in a condensed state represents a macromolecular coil which is a fractal [2]. The feature of fractal structures is the different degree of accessibility of their units in different areas of a fractal: it is much larger on the surface of fractal, than in its internal areas [3]. Therefore it is necessary to assume that the effective number of the reactive sites of a polymeric macromolecule is defined not only by its chemical constitution, but also by the macromolecular coil structure. The purpose of the present paper is a definition of influence of the macromolecular coil structure on the effective number of the macromolecule reactive centers by an example of a melt of the block copolymers of polyarylatearylenesulphonoxide (PAASO), obtained by the different methods of polycondensation [4]. For reaching this purpose the representations of the fractal analysis [5, 6] and the scaling approach [7] are used.

EXPERIMENTAL

PAASO, obtained by the different methods of polycondensation is studied. These methods and legends of block copolymers are listed in table 1. PAASO is synthesized from diane, mixture (1:1) of dichloroanhydrides of tere- and isophalic acids and 4,4'-dichlorodiphenylsulfone with molecular weight 4600, having the following structure:

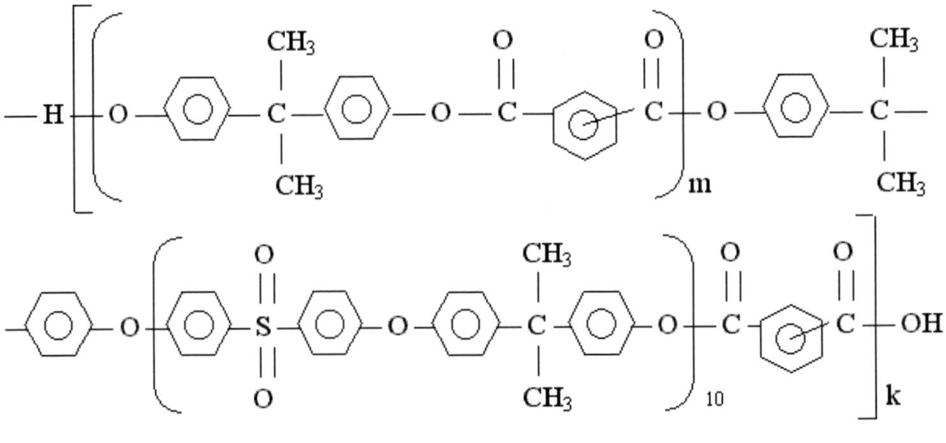

The average molecular weight $\overline{M_w}$ is determined by method of approaching to equilibrium (Archibald method) in ultracentrifuge 3170 of the corporation MOM (Hungary). The values $\overline{M_w}$ for PAASO are listed in table 1.

The glass transition temperature T_g of the studied copolymers is defined by the dielectric method. The studies are carried out by quantometer BM-560 "Tesla" at frequency 1 MHz within temperature range 293-573 K [4]. The values T_g for PAASO are also listed in table 1.

For studies of the thermooxidative degradation processes in the air the ampullary technique is used. The working volume of ampoules is equal to 3×10^{-5} *l*. The average initial contents of oxygen make up the value 2.5-3.0 mol. O_2/mol of polymer. The kinetic curve oxygen consumption $N_{O_2}(t)$ is obtained within temperature range 573-723 K [4].

The plotting of logarithmic anamorphoses of kinetic curves N_{O_2} (*t*) allows to determine the constant of the reaction rate k_d for thermooxidative degradation (see table 1).

Table 1. Methods of Polycondensation, Legends and Basic Characteristics of Studied PAASO

The method of polycondensation	A legend	T_g, K	$\overline{M_w}$ ×10^{-3}	k_d×10^{-4}*, s^{-1}
Low-temperature	PAASO-1	472	76	0,19
High-temperature	PAASO-2	474	64	0,24
Interfacial	PAASO-3	491	66	0,07
Emulsive	PAASO-4	489	58	0,11

*At *T*=623 K.

RESULTS AND DISCUSSION

In Fig. 1 two typical kinetic curves of oxygen consumption N_{O_2} (*t*) for PAASO-1 at two temperatures of testing *T*: 623 and 723 K are shown. As follows from the data of Fig. 1, these curves have two main differences: the different rates of oxygen consumption and the different values N_{O_2}, corresponding to an the outlet of a kinetic curve on a plateau of asymptotic values. Let's consider the analysis of these differences in frameworks of the scaling approach [8]. As an example we shall consider the reaction in which particles P of the chemical substance diffuse in the environment containing randomly disposed static nonsaturated traps T. At the contact of a particle P with the trap T the particle disappears. The nonsaturation of the trap means that the reaction P+T→T can be repeated an infinite number of times. It is usually considered that, if the concentration of particles and traps is great or the reaction happens at intensive stirring, the process can be considered as the classical reaction of the first order. In this case it is possible to consider that the law of decay of particle concentration with time *c*(*t*) will look like [8]:

$$c(t) \approx \exp(-At), \qquad (1)$$

where *A* is a constant, proportional to the concentration of traps (reactive centers of polymeric macromolecules).

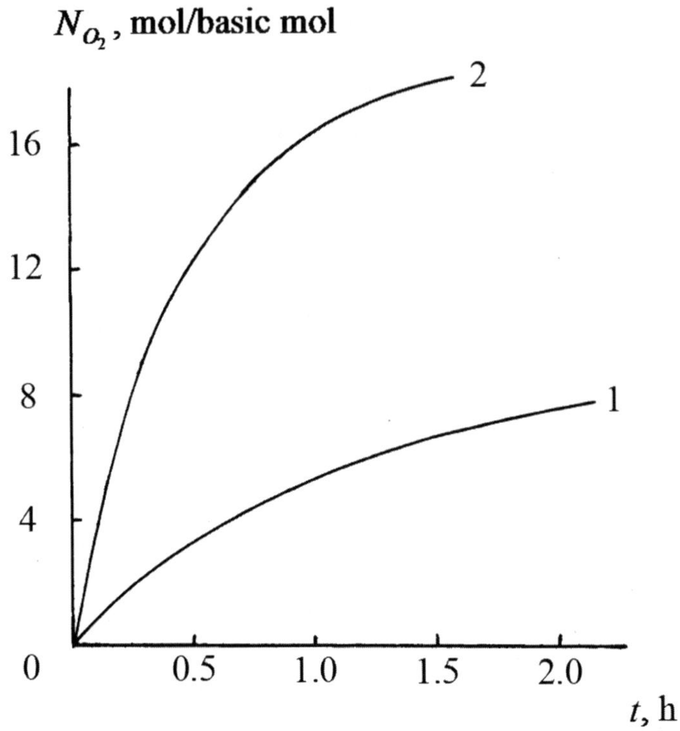

Fig. 1. Kinetic curves of oxygen consumption N_{O_2} as the function of time t for PAASO-1 at T=623 (1) and 723 K (2).

In Fig. 2 the dependences $c(t)$ according to equation (1) for curves $N_{O_2}(t)$, plotted in Fig. 1, where the value c is defined as ($N_{O_2}^{\infty} - N_{O_2}$) ($N_{O_2}^{\infty}$ - the limiting quantity of oxygen which is capable to be consumed at oxidation) are shown. The values $N_{O_2}^{\infty}$ are accepted to be equal to 12 mol/basic mol for T=623 K and 24 mol/basic mol for T=723 K [9]. As follows from the data of Fig. 2, dependences $c(t)$ plotted according to equation (1), are linear, i.e. this equation correctly describes the process of thermooxidative degradation PAASO-1. It means that in the considered case the high concentration of particles (molecules of oxygen) and traps (reactive centers of macromolecules) takes place and latter as it should have been expected, are immobile. The dependences $c(t) \approx \exp(-At)$, which are similar shown in Fig. 2, allow to determine values A for PAASO.

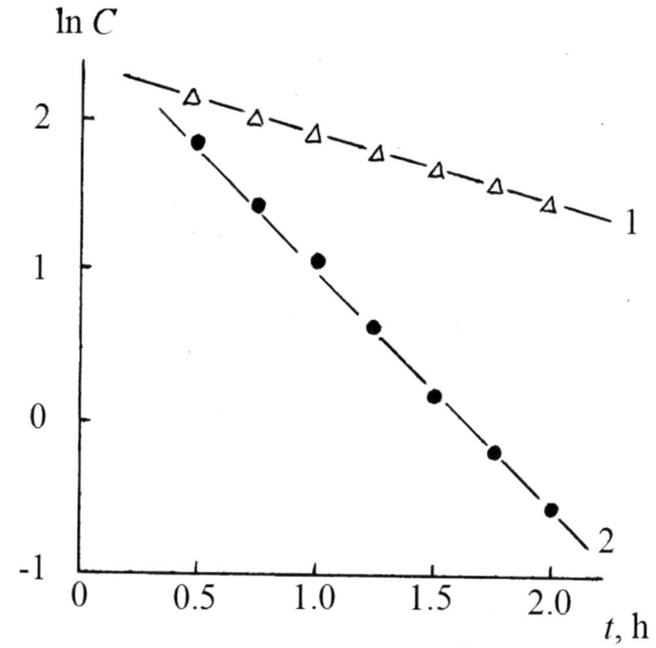

Fig. 2. Dependences of particle concentration c decay with time t in log-log coordinates according to equation (1) for PAASO-1 at T=623 (1) and 723 K (2).

Furter we shall consider the parameters describing the structure of a polymeric melt. As it was mentioned above, the study of the process of thermooxidative degradation of PAASO is carried out in temperature range 573-723 K, and the temperature range of glass transition temperatures T_g of these copolymers is T_g=472-491 K (table 1). As it is known [10], the temperature of so-called transition "liquid 1 – liquid 2" T_{ll} can be estimated as follows:

$$T_{ll} \approx (1.20 \pm 0.05) T_g . \tag{2}$$

From equation (2) and the data of table 1 the condition $T \geq T_{ll}$ follows. At T_{ll} there is a transition of a polymeric melt from "liquid with fixed structure" (where the residual structural ordering [10] is observed) to the true liquid state or "unstructurity" of a melt at $T \geq T_{ll}$ concerns the absence of the supermolecular structure, but the structure of a macromolecular coil in a melt remains by the important structural factor (essentially, unique at $T \geq T_{ll}$).

It is possible to describe structure of a macromolecular coil, which is the fractal object [2], more exactly with the help by its fractal (Hausdorff) dimension Δ_f, describing the distribution of coil units in space [12]. The estimation of the value Δ_f can be made as follows [13]. The formal kinetics of chemical reactions can be described by the following equation:

$$\frac{dQ}{dt} = k_d (1 - Q). \tag{3}$$

where Q is a reaction degree of conversion, t is its duration, k_d is a constant of the reaction rate.

The general fractal relationship also used for the description of kinetics of chemical reactions, looks like [6]:

$$Q \sim t^{(3-\Delta_f)/2}. \tag{4}$$

Differentiating relationship (4) by time t and equating the derivative dQ/dt to similar derivative in equation (3), we shall receive [13]:

$$t^{(\Delta_f-1)/2} = \frac{C_1}{k_d(1-Q)}, \tag{5}$$

where C_1 is a constant, which can be estimated from the boundary conditions [13], and the value Q is defined as a ratio $N_{O_2}/N_{O_2}^{\infty}$.

In Fig. 3 the dependence $A(\Delta_f)$ for PAASO is shown. As it should have been expected, the increase Δ_f or the rise of a degree of a macromolecular coil compactness [12] results in lowering A or decrease of the effective number of reactive centers of a macromolecular coil and at $\Delta_f \approx 2.8$ the access of oxidant in these domains is screened by its surface, i.e., in this case only those reactive centers of a macromolecule can be exposed to the oxidation which are disposed on a coil surface.

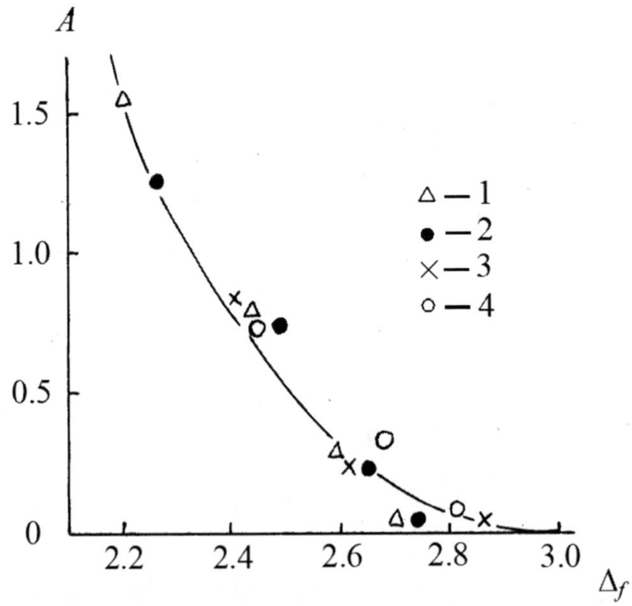

Fig. 3. Dependence of the parameter A on fractal dimension of a macromolecular coil Δ_f for PAASO-1 (1), PAASO-2, PAASO-3 (3) and PAASO-4 (4).

Thus, the process of a melt PAASO oxidation is passed in a fractal space with dimension Δ_f. In such space the process of degradation can be presented by the way of "devil's staircase" [14]. Its horizontal segments correspond to time intervals, where the reaction does not pass. In this case the degradation process is described with the usage of fractal time t, which belongs to point of Cantor's set [15]. If the reaction is considered in an Euclidean space, the time belongs to the set of real numbers.

For the description of evolutional processes with fractal time the mathematics of fractional differentiation and integration will be used [15]. As it is shown in paper [16], in this case the fractional exponent ν coincides the fractal dimension of Cantor set and indicates a fraction of the system nonchanged for all time of evolution t. We remind that the Cantor set is considered in one-dimensional Euclidean space (d=1) and consequently its fractal dimension d_f<1 by virtue of fractal definition [14]. For fractal objects in Euclidean spaces with higher dimensions (d>1) as ν it is necessary to accept the fractional part d_f (in our case Δ_f) or:

$$\nu = \Delta_f - (d-1). \tag{6}$$

Then, value ν characterizes a fraction of a fractal (macromolecular coil), unchanged during the process of degradation.

It is possible to assume that the value A or effective number of the reactive centers of a macromolecule coil being exposed to the degradation, or (1-ν). In Fig. 4 the dependence A on (1-ν) for PAASO is shown, it has appeared linear and value A at (1-ν)≈0.21 or Δ_f≈2.79 reaches approximately the constant value that is small according to the absolute magnitude: A≈0.04. This value A corresponded with the oxidant reaction only on the surface of a macromolecular coil. For Δ_f<2.80 it is possible to write the analytical relationship between A and ν:

$$A = 0.5(4-5\nu), \tag{7}$$

and then with the allowance for equation (1) we shall receive:

$$Q \approx 1-\exp\left[-0.5(4-5\nu)t\right]. \tag{8}$$

From equation (8) the follows that at ν>0.8 the mechanism of the thermooxidative degradation changes, as an exponent in this equation becomes positive.

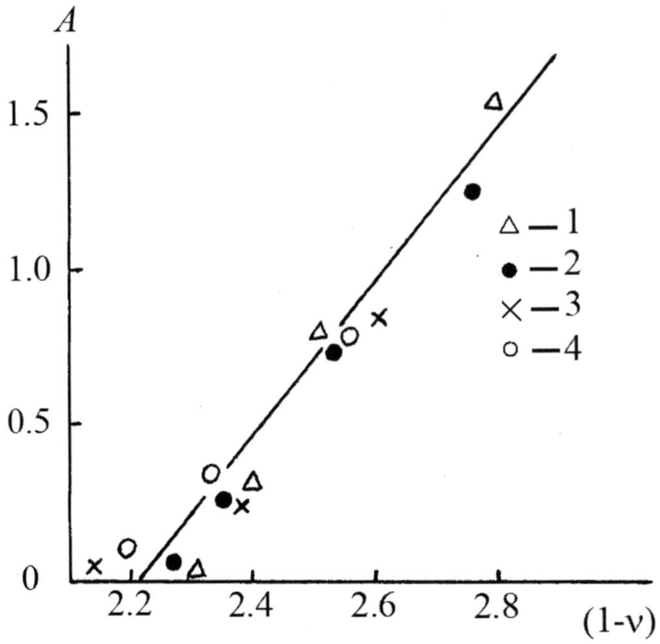

Fig. 4. Dependence of the parameter *A* on fraction (1-ν) of macromolecular coil subjecting degradation for PAASO-1 (1), PAASO-2, PAASO-3 (3) and PAASO-4 (4).

CONCLUSIONS

The results of the present paper have shown, that the effective number of the polymeric macromolecule reactive centers depends on the macromolecular coil structure. The quantitative characteristic of the latter one can be received within the framework of the fractal analysis. The increase of macromolecular coil fractal dimension meaning the rise of a degree of its compactness reduces in lowering the effective number of the reactive (accessible to the molecules of oxidant) sites of a polymeric macromolecule.

REFERENCES

[1] Shlyapnikov Yu.A., Kirjushkin S.G., Mar'in A.P. *Antioxidative stabilization of polymers*. Moscow, Khimiya, 1986, 256 p.

[2] Baranov V.G., Frenkel S.Ya., Brestkin Yu.V. *Dokl. AN SSSR*, 1986, v. 290, № 2, p. 369.

[3] Pfeifer P., Avnir D., Farin D. *J. Stat. Phys*., 1984, v. 36, № 5/6, p. 699.

[4] Batyrova H.M. *Diss.... kand. khim. nauk*. Moscow, MkhTI, 1985, 160 p.

[5] Kozlov G.V., Novikov V.U. *Synergetics and fractal analysis of cross-linked polymers*. Moscow, Klassika, 1998, 112 p.

[6] Novikov V.U., Kozlov G.V. *Uspekhi khimii*, 2000, v. 69, № 6, p. 572.

[7] De Gennes P.G. *Scaling concept in polymer physics*, Cornell U.P., Ithaca, 1979, 367 p.

[8] Djordjevič Z.B. In book: *Fractals in Physics*. North-Holland, Amsterdam, Oxford, New York, Tokyo, 1986, p. 581-585.

[9] Shabaev A.S. *Diss.... kand. khim. nauk*. Moscow, MkhTI, 1985, 106 p.

[10] Bershtein V.A., Egorov V.M. *Differential scanning calorymetry in physic-chemistry of polymers*. Leningrad, Khimiya, 1990, 256 p.

[11] Lobanov A.M., Frenkel S.Ya. *Vysokomolek. soed. A*, 1980, v. 22, № 5, p. 1045.

[12] Vilgis T.A. *Physica A*, 1988, v. 153, № 2, p. 341.

[13] Ligidov M.H., Kozlov G.V., Bejev A.A. *Izvestiya VUZov – Khimiya*, 2001, v. 44, № 3, p. 27.

[14] Halsey T.C., Jensen M.H., Kadanoff L.P., Procaccia I., Shraiman B.I. *Phys. Rev. A*, 1986, v. 33, № 2, p. 1141.

[15] Bolotov V.N. *Pisma v ZhTF*, 1995, v. 21, № 10, p. 82.

[16] Nigmatullin R.R. *Teoret. i matemat. Fizika*, 1992, v. 90, № 3, p. 354.

Chapter 4

THE SCALING ANALYSIS OF THERMOOXIDATIVE DEGRADATION OF POLYARYLATEARYLENESULFONOXIDE

G. V. Kozlov and H. M. Batyrova

Kabardino-Balkarian State University

Chernishevsky st. 173, 360004, Nalchik, Russia

G. E. Zaikov

Institute of Biochemical Physics Russian Academy of Sciences

Kosygin st. 4, 119991, Moscow, Russia

ABSTRACT

It is shown, that scaling relationships adequately describe the process of oxygen consumption in the course of thermooxidative degradation of polyarylatearylenesulfonoxide. The correspondence of the experimental curve amount of the consumed oxygen-time to this or that scaling relationship allows to define the mechanism of thermooxidative degradation. So, for polyarylatearylenesulfonoxide the mode of autodeceleration corresponds to the attack by oxygen molecules the polymeric macromolecules reactive centers, and the mode of autoacceleration correspoinds to the reaction of low molecular radicals.

INTRODUCTION

The scaling concepts are widely applied in physics of polymers [1] and it was shown [2] that they can be used for the description of chemical reactions. The scaling laws always define some asymptotics, whose applicability taking into account the specificities of the system, needs to be analyses for each concrete case [3].

As an example we shall consider the reaction in which particles P of a chemical substance diffuse in the environment containing randomly disposed static nonsaturated traps T. At the contact of particle P with trap T the particle disappears. The nonsaturation of the trap means that the reaction $P+T\rightarrow T$ can be repeated an infinite number of times. It is usually considered that if the concentration of particles and traps is great or the reaction occurs at the intensive stirring, the process can be considered as the classical reaction of the first order. In this case it is possible to consider that the law of decay of the particle concentration with time $c(t)$ will be [2]:

$$c(t) \approx \exp(-At), \tag{1}$$

where A is a constant, proportional to the concentration of traps.

However, if the concentration of randomly disposed traps is small, with the necessity there exist the areas of space which are practically free from traps. The particles which got into these areas, can reach traps only for rather big time, and, therefore, the decay of their number will be more slower. The formal analysis of this problem shows that the concentration of particles decays under the law [2]:

$$c(t) \approx \exp\left(-Bt^{d/(d+2)}\right), \tag{2}$$

that depends on the dimension of space (B is a constant).

It is necessary to note that the singular dependence on time in equation (2) occurs simultaneously with the large-scale fluctuations of trap density. If the traps can move, their motion as if averages the effect of space fluctuations, so that the suppositions resulting in to equation (1) will be fulfilled better. It was shown that in this case the concentration of particles decays under the combined law [2]:

$$c(t) \sim \exp(-At)\exp\left(-Bt^{d/(d+2)}\right), \tag{3}$$

where A is proportional to the traps diffusivity.

The purpose of the present paper is the descriptions of kinetics of the oxygen consumption within the process of a polyarylatearylenesulfonoxide (PAASO) thermooxidative degradation in frameworks of the scaling approach. PAASO, obtained by the different methods of polycondensation, are the convenient object of study, as at the identical chemical constitution they have detected the different curves amount of consumed oxygen-time $N_{O_2}(t)$ not only in quantitative, but also in qualitative relation, i.e., they have the different types of curves $N_{O_2}(t)$ – with autoacceleration and autodeceleration [4].

Experimental

We studied PAASO, obtained by the low-temperature acceptor-catalytic (PAASO-1) and interphase (PAASO-2) polycondensation. This block copolymer is synthesized from diane, mixture of (1:1) dichloroanhydrides of tere- and isophalic acids and 4,4'-dichlorodiphenylsulfone with molecular weight 4600, being of the following structure:

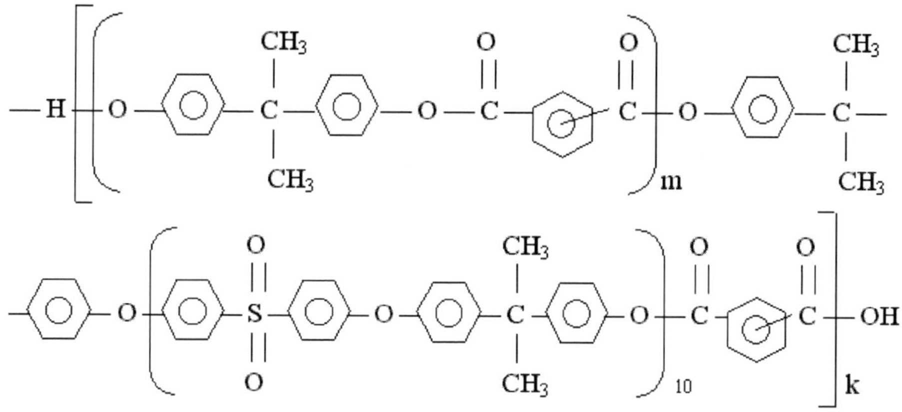

The average molecular weight $\overline{M_w}$ is determined by a method of approaching to equilibrium (Archibald method) in ultracentrifuge 3170 of corporation MOM (Hungary). The values $\overline{M_w}$ are equal to 76×10^3 for PAASO-1 and 69×10^3 for PAASO-2 [5]. The glass transition temperature T_g of studied copolymers is defined by a dielectric method. The studies are carried out by quantometer BM-560 "Tesla" at frequency 1 MHz in temperature range 293-573 K [5]. The value T_g is equal to 471 K for PAASO-1 and 491 K – for PAASO-2.

For the studies of thermooxidative degradation processes in air the ampullary technique is used. The working volume of ampoules is equal to 3×10^{-5} *l*. The average initial contents of oxygen make up the value 2.5-3.0 mol. O_2/mol of polymer. The kinetic curves of the oxygen consumption $N_{O_2}(t)$ is obtained at temperatures 623 and 723 K.

RESULTS AND DISCUSSION

In Fig. 1 three studied kinetic curves $N_{O_2}(t)$ for PAASO are shown. As it is possible to see, at T=623 K the curves $N_{O_2}(t)$ for these copolymers differ by the type: for PAASO-1 the curve $N_{O_2}(t)$ has the autodecelerated character, and for PAASO-2 – autoaccelerated. The curve $N_{O_2}(t)$ for PAASO-1 at T=723 K also has the autodecelerated character, but with much bigger gradient of change N_{O_2} in accordance with the increase of time t. Let's consider the scaling treatment of the shown in Fig. 1 curves $N_{O_2}(t)$ with the usage of equations (1)-(3) [2]. In Fig. 2 the dependences $c(t)$ according to equations (1) and (2) for PAASO-1 are shown at T=623 K (mode of autodeceleration), where the value c is defined as ($N_{O_2}^{\infty}$ - N_{O_2}) ($N_{O_2}^{\infty}$ is limitary quantity of oxygen which is capable to be consumed at oxidation). The values $N_{O_2}^{\infty}$ are accepted to be over equal to 12 mol/ basic mol for T=623 K and 24 mol/basic mol for 723 K [6]. As follows from the data in Fig. 2 the dependences $c(t)$, drawn according to both mentioned equations, are linear, i.e., these equations correctly describe the thermooxidative degradation process of PAASO-1. It means that in the considered case the intermediate concentration of particles (molecules of oxygen) and traps (reactive centers of polymeric macromolecules) takes place, and the latter, as well as it was necessary to expect, are immobile. In Fig. 3 the dependences $c(t)$, drawn according to equations (1)-(3) for PAASO-1 are shown at T=723 K. As against the graphs of Fig. 2 for the same copolymer, but at lower temperature, none of the indicated equations describes the experimental data. As it is shown in paper [7] on an example of cross-linking reaction of epoxy polymers, it means, that the reaction takes place not in an Euclidean, but in a fractal space with dimension d_f. Proceeding from this, it is necessary to consider the physical sense of this fractal dimension. As it was mentioned above, the study of the thermooxidative degradation process of PAASO is made within the temperature range 623-723 K and the temperature range of glass transition of these copolymers is equal to 471-491 K. As it is known [8], the temperature of so-called transition "a liquid 1 – liquid 2" T_{ll} can be estimated as follows:

$$T_{ll} \approx \left(1.20 \pm 0.05\right) T_g . \tag{4}$$

From equation (4) and quoted above data the condition $T>T_{ll}$ follows. At T_{ll} there occurs a transition of polymeric melt from "liquid with fixed structure" (where the residial structural ordering is observed [8]) to true liquid state or "unstructured liquid" [9]. Nevertheless, "unstructurity" of melt at $T>T_{ll}$ relates to the absence of supermolecular structure, but the structure of a macromolecular coil in melt remains the important structural factor (in essence, unique at $T>T_{ll}$).

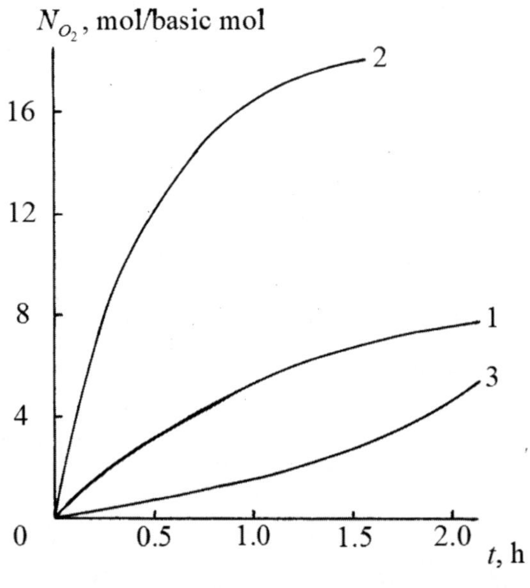

Fig. 1. Kinetic curves of oxygen consumed amount N_{O_2} as the function of time t for PAASO-1 at T=623 (1) and 723 K (2) and PAASO-2 at T=623 K (3).

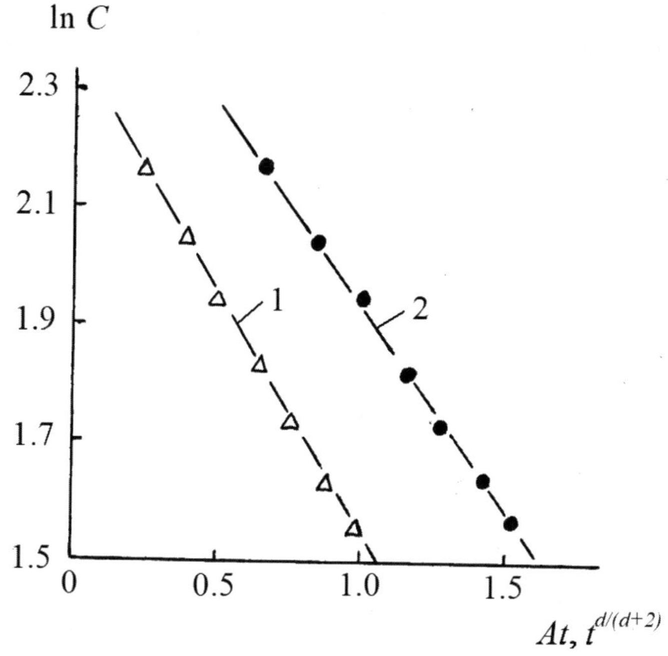

Fig. 2. Dependences of decay of concentration of particles c on time t in logarithmic coordinates according to equations (1) (1), (2) (2) and (3) (3) for PAASO-1 at T=623 K.

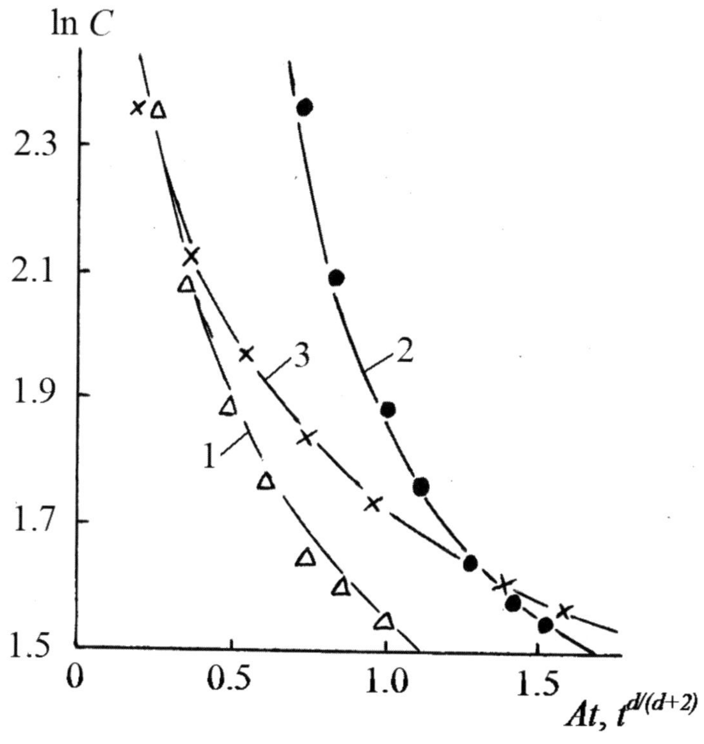

Fig. 3. Dependences of decay of concentration of particles c on time t in logarithmic coordinates according to equations (1) (1), (2) (2) and (3) (3) for PAASO-1 at T=723 K.

The structure of a macromolecular coil, which is fractal object [10], may be precisely described with the help of its fractal (Hausdorff) dimension Δ_f, describing the distribution of units of a coil in space [3]. Therefore, in the considered case, the fractal space of reaction is formed by the fractal structure of a macromolecular coil and consequently $d_f=\Delta_f$. Further, in an exponent of the right part of equation (2) it is necessary to replace the Euclidean dimension d with the fractal dimension Δ_f. In Fig. 4 the dependence $c(t)$ for PAASO-1 is shown at T=723 K, corresponding to equation (2) with the mentioned above replacement. As it is possible to see in this case, the linear relationship ln c as the function t is obtained, i.e., the process of oxidation realizes at small concentration of oxygen molecules in the fractal space. At first sight, the temperature rise from 623 K up to 723 K causes the change of the reactionary environment character from the Euclidean to the fractal. More in-depth study has shown that the problem consists of the value Δ_f or, more precisely, of the closeness of dimension Δ_f to d, that determines the closeness of numerical values of an exponent in equation (2) at usage Δ_f and d. So, for PAASO-1 at T=623 K $\Delta_f\approx2.6$ and at T=723 K $\Delta_f\approx2.2$ [11]. Correspondly, in the first case the difference of values $t^{d/(d+2)}$ and $t^{\Delta_f/(\Delta_f+2)}$ does not exceed 2.5 %, and in the second this difference there are more than 5%. Therefore, the drawn of dependence ln c ($t^{\Delta_f/(\Delta_f+2)}$) for PAASO-1 at T=623 K has shown that these dependences for Euclidean (d=3) and fractal (d=2.6) spaces are practically described by one linear correlation. For PAASO-1 at T=723 K the difference for cases d=3 and Δ_f=2.2 is already apparent, that follows from comparing the graphs of Fig. 3 and 4, drawn according to equation (2). And a last, in Fig. 5 the dependences $c(t)$ drawn according to equations (1)-(3) for PAASO-2 are shown at T=623 K, i.e., for the mode of autoacceleration. From the pointed graphs follows that equations (1) and (2) do not describe the behaviour PAASO-2 in the mode of autoacceleration, whereas the linearity of dependence $c(t)$, drawn according to equation (3), shows the correctness in this case. It means that the traps can move, like the particles. Therefore, as in the mode of autoacceleration the thermooxidative degradation realizes at the expense of radical reaction [12], their high mobility supposes that in case of PAASO the low molecular radicals react. The probable chemical mechanism of formation of such radicals in case of high-temperature oxidation is discussed in [12]. It is supposed that the product, responsible for autoacceleration, is the aldehydic groups, formed from the methyl groups according the scheme [12]:

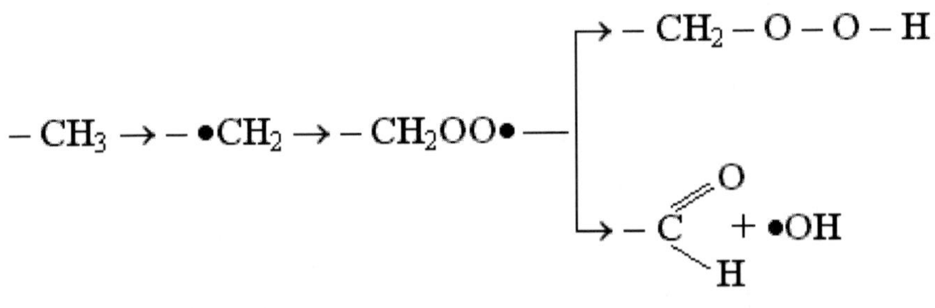

The oxidation of the aldehydic groups realizing according the scheme [12]

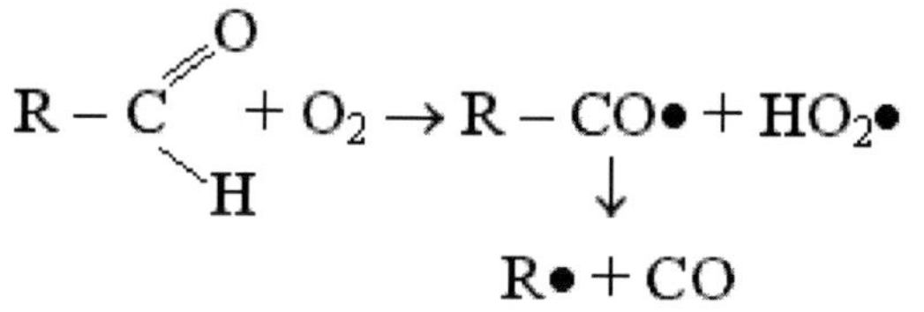

results in the formation of active free radicals R• and low molecular radicals HO_2• [12].

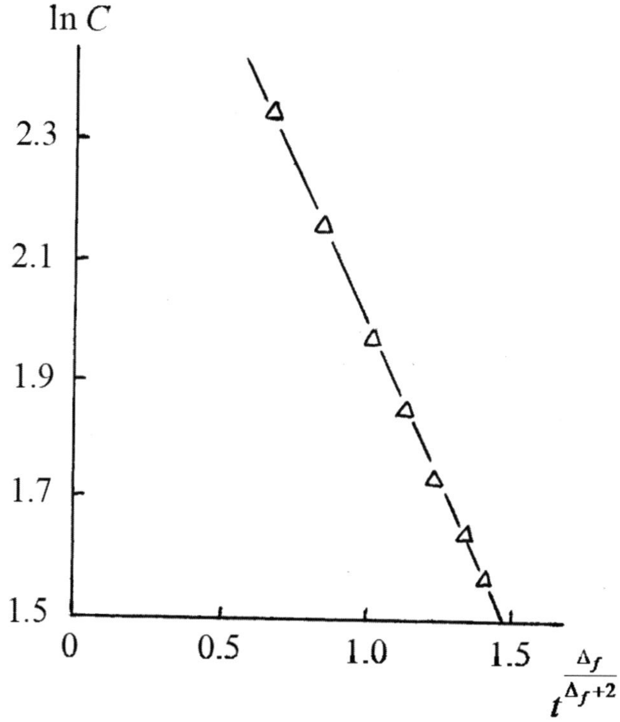

Fig. 4. Dependences of decay of concentration of particles *c* on time *t* according to equation (3), in a fractal space with dimension Δ_f for PAASO-1 at *T*=723 K.

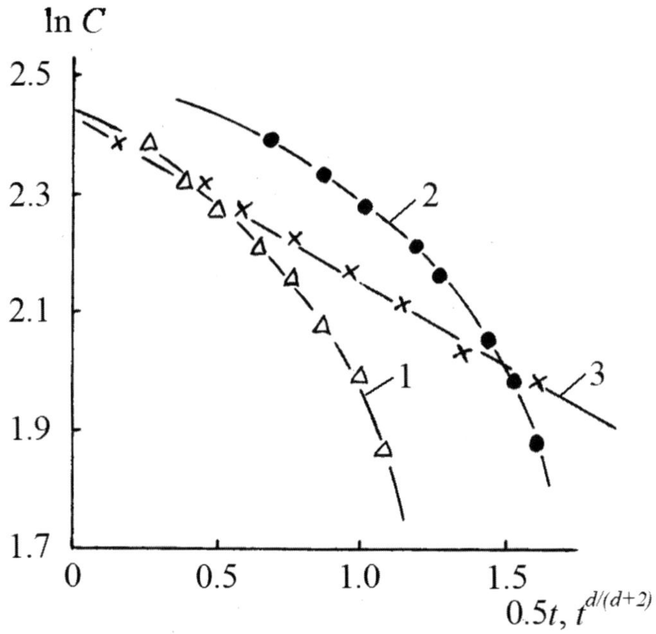

Fig. 5. Dependences of decay of concentration of particles *c* on time *t* according to equations (1) (1), (2) (2) and (3) (3) for PAASO-2 at *T*=623 K in logarithmic coordinates.

Conclusions

Thus, the results, obtained in the present paper, have shown, that scaling relationships adequately describe the process of oxygen consumption in the course of thermooxidative degradation of polyarylatearylenesulfonoxide. The correspondence of the experimental curve amount of the consumed oxygen-time to this or that scaling relationship allows to define the mechanism of thermooxidative degradation. So, for polyarylatearylenesulfonoxide the mode of autodeceleration corresponds to the attack by oxygen molecules the polymeric macromolecules reactive centers, and the mode of autoacceleration correspoinds to the reaction of low molecular radicals.

References

[1] De Gennes P.G. *Scaing Concepts in Polymer Physics*, Cornell Univ. Press, 1979, 342 p.

[2] Djordjevič Z.B. In book: *Fractals in Physics*. Ed. Pietronero L., Tosatti E. Amsterdam-Oxford-New York-Tokyo, North-Holland, 1986, p. 581-585.

[3] Kokorevich A.G., Gravitis Ya.A., Osol-Kalnin V.G. *Khimija drevesiny*, 1989, № 1, p. 3-24.

[4] Emanuel N.M. *Vysokomolek. soed. A*, 1985, v. 27, № 7, p. 1347-1363.

[5] Batyrova H.M. *Diss. ... kand. khim. nauk*, Moscow, MkhTI, 1985, 160 p.

[6] Shabaev A.S. *Diss. ... kand. khim. nauk*, Moscow, MkhTI, 1985, 109 p.

[7] Kozlov G.V., Bejev A.A., Shustov G.B., Lipatov Yu.S. *Intern. Conf. "Olygomers-2000"*. 4-8 Sept. 2000, Perm. Abstracts, p. 207.

[8] Bershtein V.A., Egorov V.M. *Differential scanning calorymetry in physic-chemistry of polymers*. Leningrad, Khimija, 1990, 256 p.

[9] Lobanov A.M., Frenkel S.Ya. *Vysokomolek. soed. A*, 1980, v. 22, № 5, p. 1045-1057.

[10] Baranov V.G., Frenkel S.Ya., Brestkin Yu.V. *Dokl. AN SSSR*, 1986, v. 290, № 2, p. 369-372.

[11] Kozlov G.V., Dolbin I.V., Batyrova H.M., Zaikov G.E. *Mater. III Al-Russian Conf. "New chemical techniques: production and application"*. Pensa, 2001, p. 61-63.

[12] Shlyapnikov Yu.A., Kiryushkin S.G., Mar'in A.P. *The oxidative stabilization of polymers*. Moscow, Khimija, 1986, 256 p.

Chapter 5

STAGES OF PROCESSES OF EXTENSION DEFORMATION OF CROSS-LINKED POLYMER OBTAINED BY HARDENING RESORCINE GLYCIDIL ESTER BY METAPHENYLENEDIAMINE AT TEMPERATURES ABOVE GLASS TEMPERATURE

E. F. Vainstein[1], *E. S. Solodysheva*[2] *and G. E. Zaikov*[1]

1. N.M. Emanuel Institute of Biochemical Physics, Russian Academy of Sciences
4 Kosygin Street, Moscow 117334, Russia

2. N.N. Semenov Institute of Chemical Physics, Russian Academy of Sciences
4 Kosygin Street, Moscow 117334, Russia

An opportunity to evaluate deformation stages was considered studying one-axis extension deformation of moderately cross-linked epoxy polymers at constant load or constant rate of extension. Kinetic equations describing the processes are written using variation of thermodynamic potentials, particularly of free energy. It is shown that deformation passes three reversible processes step-by-step: lining-up of non-associated parts of chain (fast stage), breakdown of intermolecular bonds with subsequent deformations (slow stage) and relative displacement of separate parts of the network (the slowest stage).

Key words: epoxy polymer networks, one-axis deformation, reversibility of kinetic processes, deformation stages, constant load, constant rate, intermolecular interaction, non-associated parts, elastic deformation

Physico-chemical parameters of deformation of cross-linked three-dimensional polymers are normally studied using Maxwell-Gurevich-Alexandrov equation [1]. Equation parameters which are deduced from experimental data either in case of constant rate of extension (v_ε = const) or constant load (σ = const) depends on polymer structure [1-3]. A physico-chemical meaning of this dependence is unclear.

It was shown experimentally that curves of extension and compression obtained for polymers under the same external conditions (temperature, ν_ε, one-axis load) can be different even qualitatively [3]. It can be assumed that in general case a deformation process is mote complicated than postulated by Maxwell-Gurevich-Alexandrov equation. Probably, cases cannot be excluded when deformation process passes fewer stages. If multistage deformation process is assumed and several structural elements are altered throughout deformation process, then to realize a link of structure to physico-mechanical properties of samples it is necessary to know the stage at which experimental data is obtained. This is apart from information about sample structure.

Since deformation is kinetic process, the kinetic method should be one of the mains to evaluating deformation stages. For this reason this work is aiming in a development such a kinetic method. There is considered the case of one-axis extension under constant lead or at constant rate of deformation. Experiments were performed at temperatures above T_g of polymers.

Three samples of epoxy polymers obtained by hardening purified resorcine diglycedyl ester and methaphenylene diamine (RDGE:MPDA = 1:1) were studied under constant load. Temperature of sample deformation at constant rate was 413 K.

The technique of sample preparation is described in details in Ref. [3]. Here we emphasize only the use of the mould with teflon inner part of which the shape and the size was defined by the size of the sample to be treated (diameter (d) = 5 mm, length (l) = 40 mm). Hardening conditions and procedure of thermal pre-treatment were similar to those described in Ref. [3, 4]. The properties of samples remained unchanged during investigation of physico-mechanical properties; these properties did not also depend on the temperature of pre-treatment.

Sample creep (σ = const) was studied at temperatures which were 20-30°C above T_g. Creep stress (σ = 0.5 kg/mm^2) was 3.5% of maximum value obtained in experiments at constant rate of deformation and at room temperature. To provide constant stress the technique was used which was earlier described in Ref. [5]. It was similar to that described in Ref. [6]. The fixed sample was loaded instantaneously (0.5-1 sec) after stabilization period of 10-15 min at constant temperature. Deformation was monitored using catetometer [5].

Experimental curves of creep are shown in Fig. 1a for DGER:MPDA samples at different temperatures of deformation. The curve of creep of linear polymer [1] is shown in Fig. 1b for comparison. Extension diagram of the sample at ν_ε = 1.0%/min is shown in Fig. 2.

Creep curves are substantially different for linear and cross-linked polymers. For instance, in case of cross-linked polymer creep increases with time approaching a limit whereas in case of linear polymer a dependence of deformation on time is more complicated. Qualitative difference in dependence of creep on time enables to use this dependence in order to realize whether a polymer is linear or cross-linked. Both curves are representative of samples which cannot be degraded or destroyed under experimental conditions.

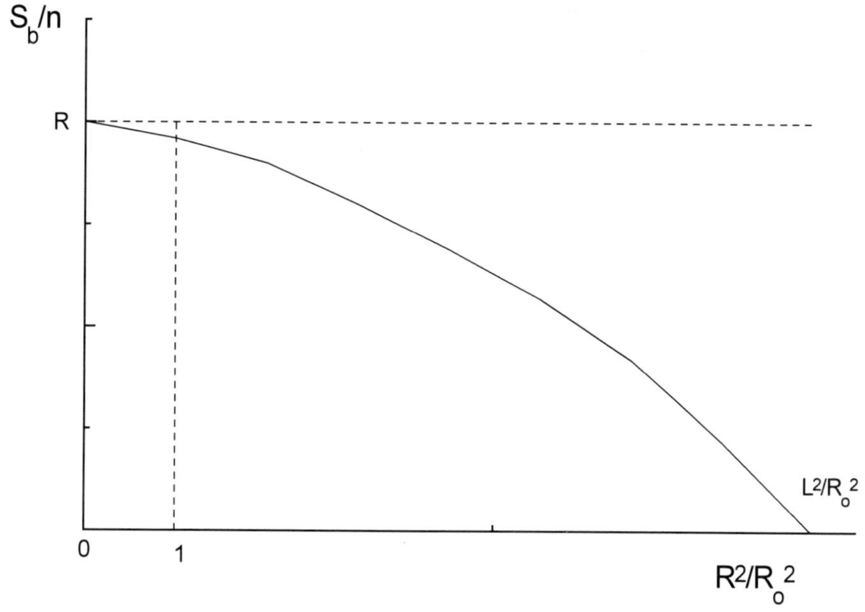

Fig. 1. Creep curves of RDGE:MPDA (a) and linear polymers (b) at different temperatures. 1- 425 K, 2- 430 K, 3 – 437 K, 4 – 443 K, 5 – 453 K. $\sigma = 0.5$ kg/cm^2

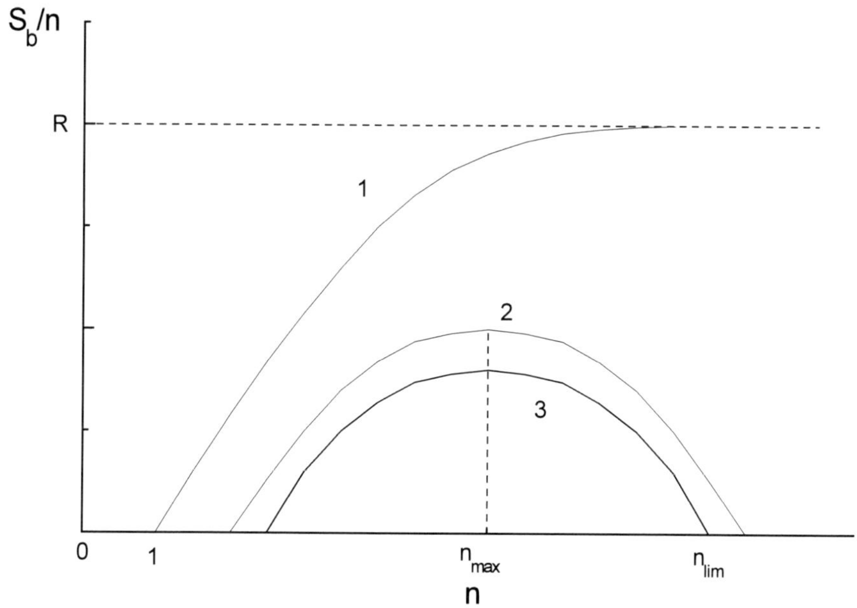

Fig. 2. Diagram of tension of crosslinked RDGE:MPDA at 413 K and $\nu_\varepsilon = 1.0$%/min

In contrast to linear polymers creep curves of cross-linked polymers are reversible, i.e. we cancel the load, heat the sample for short time and load the sample again, the resulting creep curve will be the same as in the first case. If creep process is reversible and consists of several stages, each stage is reversible. Reversibility of stage (process is a consequence of counter processes, extension and relaxation.

We presume that irreversibility of creep curves of linear polymer links to the displacements of separate molecules, which are further fixed by intermolecular bonds. For cross-linked systems substantial displacement of separate units is nearly impossible without breakdown of cross-links. The displacement of separate units in cross-linked systems (in this case new thermodynamic stationary state appears[1]) is defined by distribution of the chain units between chemical cross-links. Such displacement, which is provoked by a load, increases with increase of unit size (decrease of cross-linking degree). Irreversible degree is only possible for low cross-linked networks.

All samples investigated in this paper can be referred to moderately cross-linked networks because they are not characterized by irreversible displacement of the units.

Different final stages of the process occurring in linear and cross-linked polymers do not necessary mean the difference in initial stages of the process. Therefore, qualitative difference of deformation curves apparently indicates a complex (multistage) character of deformation process.

Comparing one-axis deformation of cross-linked polymers at constant rate (Fig. 2) or under constant load (Fig. 1) one can note that if ν_ε = const, degradation and destruction of a sample is defined by the experimental time rather than by experimental rate. Whereas under constant load, the degradation and destruction is defined by a load and consequently by deformation time. Obviously, reversibility can be observed at constant deformation rate only before degradation. Since deformation conditions are different (ν_ε = const, σ = const) in each case, kinetic equations describing processes will be different too.

One can think that stages of processes occurring at different load conditions are the same although the existence of specific studies cannot be excluded. Let us describe one-axis deformation at different deformation conditions.

Deformation under Constant Load (σ = const)

Reversibility of deformation is determined by an increase of resistance of strength force F_s to tensile force F_t. Once deformation is complete (equilibrium at fixed tensile force or at thermodynamically stationary state), strength will be equal to tensile force. Hence, the driving force of the process is resulting force R, which is equal to the difference between tensile force and strength force, i.e.

$$R = F_t - F_s \tag{1}$$

Strength force of a network is a function of free energy of extension. In the first approximation, free energy of network formation G is equal to

$$G = G_C + G_M + G_B + G_S \quad (2)$$

where G_C is the free energy of network formation of definite chemical structure neglecting unit bending, i.e. free energy taking into account all chemical groups of the network (including end-groups), G_M is the free energy of formation of intermolecular bonds (intermolecular interaction) which is equal to the sum of free energies of several intermolecular bonds, G_B is the free energy of unit bending between chemical cross-links, G_S is the free energy due to network (system) formation. The latter particularly characterizes the location of specific parts of the network in a space.

Additivity of contributing factors is due to the basic postulate of thermodynamics telling that variation of free energy does not depend on a way from initial to final stages, once initial stage is fixed [8]. A sequence of stages can be chosen, which provides the definite link between a stage and a value of free energy.

Obviously, in specific case a number of contributing factors can be more or less than that indicated in equation (2). For example, a number of contributing factors is more in cases of conformation composition or crystal phase. A number of contributing factors will be less in case of very developed networks.

The dependence of G on a sample length l is usually unknown because chain is not specified. For this reason, the above dependence is searched as a number in respect to l:

$$G = a + bl + cl^2 + dl^3 + el^4$$

Since G is finite variable, and $l < 1$, the coefficients of this number should decrease starting from definite member [9]. In the first approximation for the case of deformation with constant load, we limit a number by the first three members. In many cases it is sufficient to describe experimental data. Then following the definition of a force [10] we write

$$F_y = -|b| + 2|c|l$$

$$R = F_p + |b| - 2|c|l \quad (3)$$

Where |b| and |c| are absolute values b and c.

Following the above assumptions

$$|b| = 2|c|l_o \quad (F_y = 0)$$

$$F_p = 2|c|l_o\varepsilon \quad (F_y = F_p)$$

where l_o and l_p are initial and maximal length of samples at break, $\varepsilon = (l_p - l_o)/l_o$ is limiting relative deformation.

Similarly to the second law of mechanics we write

[1] New thermodynamic stationary state means the state of a system (sample) which is characterized by less minimum of free energy of its formation than minimum of free energy of the equilibrium. It is characterized by an existence and distribution of irreversible associates in a sample [7].

$$R = F_p + |b| - 2|c|l - md^2t/dl^2 \qquad (4)$$

The sign "-" shows the opposite sign of the increase of sample length and acceleration; m is the measure of inertia due to the resistance to tensile stress.

If a measure of inertia of a process exists, the process starts from the certain value of force sufficient to overlap measure of inertia. Apparently, this force is relatively small in comparison with the experimental force.

The solution of equation (4) is well described in literature [9-11]. If m = const, the process velocity is equal:

$$dl/dt = (F_p/(l_p - l_o)m)^{0.5}\,(l_p - l) = k(l_p - l) \qquad (5)$$

The constant of integration is defined from dl/dt = 0.

Equation (5) enables understanding of physical meaning of m and shows the link between kinetic (k) and thermodynamic (m, l_p and l_o) parameters of the process and tensile force F_p. Kinetic and thermodynamic parameters are defined by external factrs and network structure. In this case

$$k = (F_p/(l_p - l_o)m)^{0.5}$$

$$m = = (F_p/(l_p - l_o)k^2$$

Following the above assumptions l_p is defined by coefficient c only.

The rate of variation of chain length l (equation 5) is the equation of one-stage 1-st order reversible reaction. Its solution is

$$\varepsilon = \varepsilon_\infty\,[1 - \exp(-kt)]$$

where $\varepsilon = (l - l_o)/l_o$ is relative sample deformation at time t.

An attempt to describe experimental data with this equation was unsuccessful. Therefore, deformation is more complicated process or process has more than one stage.

Taking into account multicomponent character of G (equation 2) it was assumed that deformation is multistage process and each stage follows the above assumptions. Since stages are characterized by different rates, only one stage must be observed at the end of the process (all other stages are complete). Taking into account additivity of components, current relative deformation ε_c can be written in form

$$\varepsilon_c = \sum_i \varepsilon_i + \varepsilon_f\,(1 - \exp(-k\,(t - t_f)) \qquad (6)$$

where ε_c is the maximal relative deformation of i-th stage, ε_f is the maximal relative deformation at the final stage, t_f is initial time of the final stage of deformation. If final stage starts at the beginning of the process, then $t_f = 0$.

Equation (6) can be re-written in form

$$\varepsilon_o - \varepsilon_t = \omega \exp(-kt) \quad (7)$$

where ε_o is the limiting relative deformation equal to the sum of maximal relative deformations of each stage

$$\varepsilon_o = \sum_i \varepsilon_i + \varepsilon_f$$

$$\omega = \varepsilon_f (1 - \exp(kt_f))$$

Fig. 3 shows experimental curve fitting with equation (7). Although curves are qualitatively different, at the end of the process experimental data are in agreement with equation (7) for all samples. Therefore, experimental data do not contradict an assumption about several stages of the process. One stage is in agreement with assumption about proportionality between rate of elongation and difference of maximal and current sample length. For all experimental conditions one stage is nearly instantaneous. For two samples slow stage of deformation is observed. All stages are subsequent.

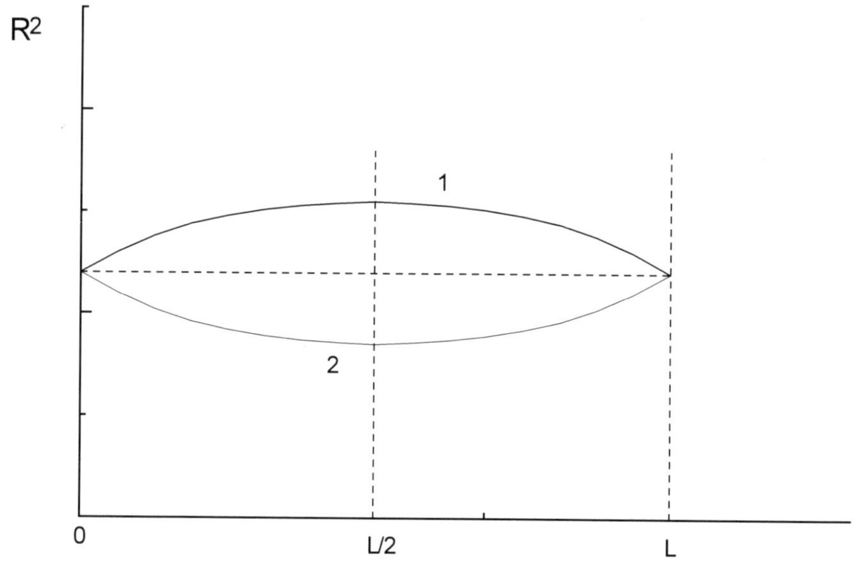

Fig. 3. Dependence of ln (ε_o - ε_t) on time for RDGE:MPDA samples at 425 K (a), 443 K (b) and 453 K (c).

Hence, three stages of are observed in conditions of deformation under constant load: one stage is fast and two are slow. All stages are reversible.

Apparently, deformation stages are described by three components of free energy (chemical structure remains unchanged because deformation is reversible). Since components of free energy have different absolute values, they are characterized by different rates. If rates are substantially different, the process is observed as a sequence of stages.

The first fast stage is apparently tension of non-associated chain units. It is of entropy character because number of stages of non-associated units is changed due to the tension. Probably, it is described by kinetic equation of reversible first order reaction. However, since deformation is faster than measurement of experimental data, only one parameter (thermodynamic) can be measured. This parameter must be linear in respect to load, if other conditions (e.g. temperature) are constant. It is true, if load does not alter polymer structure.

The second stage apparently relates to the breakdown of molecular bonds in a sample. New non-associated units are deformed instantaneously. Hence this stage is not trivial. The breakdown of intermolecular bond is limiting and defines its kinetics. Apparently, due to requirement of maximum of work [10] breakdown of intermolecular bonds starts from the end of associated units. Association of tensile units is apparently subject of the third stage. It is necessary to note that non-associated and associated units can be located within the same interval depending on unit length and intervals between chemical links. Since number and length reduce with increase of temperature, the second stage vanishes. Naturally, they are not all intermolecular bonds which are broken depending in load and external conditions. Intermolecular bonds remain in large associates.

Probably both stages are observed in conditions of one-axis deformation of linear polymers.

The third stage of the process is apparently links to the displacement of non-associated units limited by chemical bonds. Motion of those associates is also observed at this stage, which cannot be separated (irreversible associates) [7]. This stage is not trivial because new intermolecular bonds are formed due to deformation and displacement of units. Probably, a displacement of units limits the stage.

An existence of irreversible associates is not desirable in a number of processes because they play no role in processes just moving in a space.

An absence of chemical links in linear polymers is the reason for qualitative difference in deformation between crosslinked and linear polymers. The length of unit and size of chemical links define the value of displacement of units in crosslinked polymers. For this reason if distance between chemical links are long enough. The processes can be observed which are similar to those in linear polymers. If experimental data fitting requires use of higher order of magnitude of polynomial dependence G (l), equations will not be in agreement with reversible first order reactions. In this case thermodynamic functions are to be used to describe kinetics of the process.

A load can be presented as the sum of partial loads of each stage of process. The tensile force is constant, and it can be deduced from experimental data. The link between velocity and acceleration can be found using corresponding equations:

$$A = (v/k - F_p)/m$$

Equation shows that a and v are not equal to zero at the same time. Since system is reversible it oscillates near state of equilibrium. The cases of such an oscillation can be evaluated solving equation in respect to force and acceleration [9, 11]. Obviously, in general case the possibility of oscillation cannot be realized.

Oscillation indicates a validity of thermodynamic approach to writing kinetic equations of processes including mechanical processes. It is due to dynamic equilibrium, which links to interaction between counteracting processes [12]. In our case, it is the process of tension of chain units under load and the process of compression of these units due to the force of strength. The process of compression is the subject of relaxation. Apparently, relaxation is multistage process, whereas oscillation is possible only at final stage of the process because other stages do not satisfy the condition of zero velocity.

LOAD AT CONSTANT RATE (v_ε = CONST)

The same assumptions are used in this case as in the case of deformation under constant load.

Instantaneous resulting force R should be equal to zero (see equation 4) because measure of inertia is not equal to zero. Acceleration is also should be equal to zero.

An instantaneous tensile force F_p is the sum of forces, of which one F_c is required to maintain the rate of process, and another should be opposite to the force of strength

$$F_p = F_c + F_s$$

At the beginning of process $F_p = F_c + F_s$. Apparently F_c is also the function of the length of tensile chain. Since force of strength increases with increase of relative deformation, the following condition is required to maintain constant force R = 0:

$$dF_s/dt = d(F_p - F_c)/dt$$

Fig. 4 shows the dependence of derivative of force of strength on deformation. This dependence is divided in three parts.

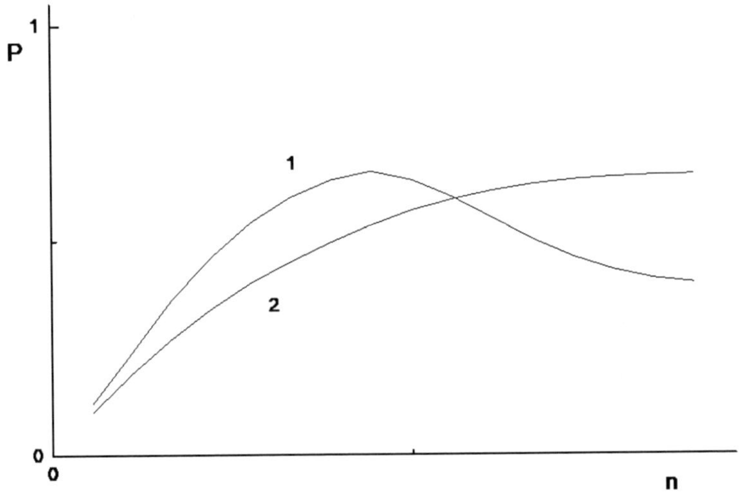

Fig. 4. Dependence of stress derivative in respect to strain against strain.

In the first part dσ/dε is constant, i.e. Hook law is valid. It is quite possible that first stage is the same for deformation under constant load and at constant rate. The weak dependence of derivative on deformation is probably due to the dependence of the rate of approach to equilibrium on deformation. Because of reversibility of each stage of the process relaxation time increases with increase of tension of a sample.

At the third stage stress does not increase and even reduces at the end of the process. This is because of displacement of separate units and absence of deformation of non-associated parts. Intermolecular bonds, which are formed in tensile and oriented pats, weakly effect on deformation and on the stress. The effect of stress (tensile force) on the energy of inter molecular bond [13] vanishes if chain units are located in associate along the force (Fig. 5). This stage is observed before destruction processes. Fig. 4 shows partial overlapping of the second and the third stages.

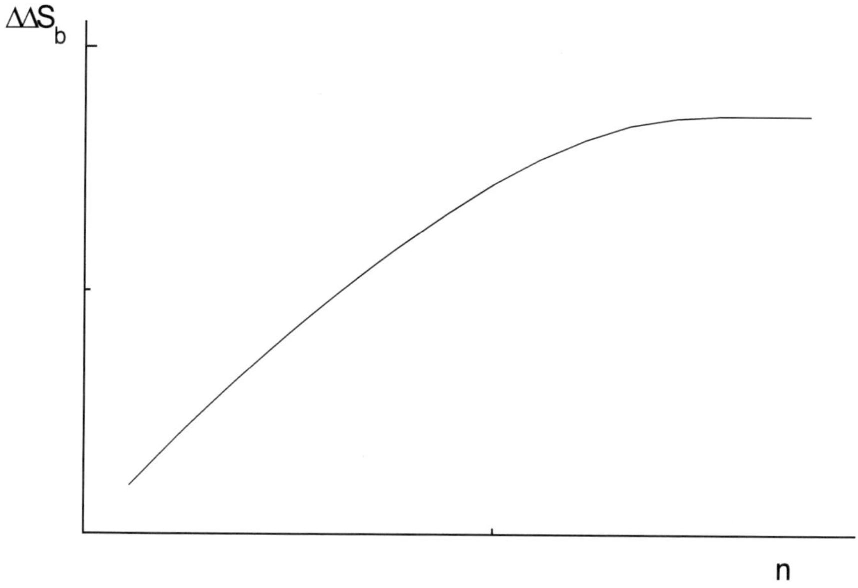

Fig. 5. Dependence of partial load $\chi = p_c/p$ on angle Θ between directions of link and load. p_c is the force provoking breakdown of intermolecular bonds.
a - Θ = 0, b - Θ = 90°, c – 0 < Θ < 90°

The dependence of stress on strain becomes weaker at the second stage (derivative decreases towards zero). Following data on creep (see above), this stage vanishes with increase of temperature due to decrease of an amount of intermolecular bonds. Apparently, this stage begins with the breakdown of intermolecular bonds and formation of longer non-associated units. The scheme of alteration of lengths of non-associated units is shown in Fig. 6. Associated block is continuous sequence of associated chain units. The blocks, which contain one intermolecular bond breaks with higher probability with decrease of the size of associated blocks that is due to the break of chain ends. This sequence of breaks is in agreement with Hamilton Principle of Minimum [10]. At the end of the second stage, mostly isolated intermolecular bonds (bonds containing one intermolecular bond) exist. For this reason their breakdown leads of the increase of relaxation time. Relaxation, which is observed at different stages (at least at first and third), is due to the reversibility of stages and process as a whole.

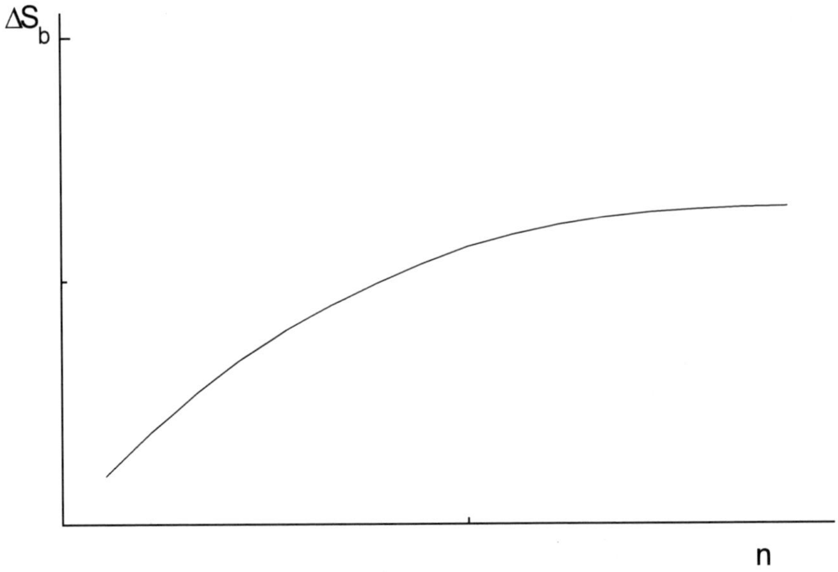

Fig. 6. The scheme of alteration of lengths of non-associated units by breakdown of intermolecular bond.
a – intermolecular bond is between non-associated parts,
b – intermolecular bond is between associated and non-associated parts.

Since mechanism of different stages of deformation is substantially different, the information can be obtained about chain length distribution between chemical links (third stage), distribution of the length of associated parts of the chain (second stage) and distribution of the length of non-associated parts of the chain (first and second stages). Obviously, correlation should be made between the above distributions.

Hence, one can hope that analysis of mechanical parameters of stages enable estimation of thermodynamics stationary states of polymer networks.

Therefore dependence shown in Fig. 4 is, at least, not in contradiction with existence of three stages and their sequence.

Apparently the main part of the stress links to the loss of entropy component of the free energy (tension of non-associated parts of the chain and their displacement. Thermal component (enthalpy) of free energy links to the breakdown (second stage) and formation (third stage) of free energy. Since stress variation is relative large and entropy of formation of intermolecular bonds is small [13], each stage is mostly of entropy character.

Due to continuous character of deformation equilibrium cannot be reached and oscillation cannot be observed. Probably, oscillation can appear when deformation is stopped and this could be a reason for dispersed character of experimental data.

Components of free energy link to molecular weight distribution of the chain length (which is determined by synthesis of a sample) and to thermodynamic stationary state (which is determine by synthesis, processing, storage and service conditions of a sample) [14]. For this reason, estimation of the effect of distribution and kind of thermodynamic stationary state on stage parameters is one of the majors.

It is unlikely that number of stages depends on molecular weight distribution of chain intervals [14].

Apparently, thermodynamic stationary state defines a number of stages. For example, increase of temperature of treatment of the same sample (number of irreversible associates vanishes) resulted in two-stage process (see Fig. 3) (namely, there were the first and third stages). The absence of the second stage at elevated temperatures indicates the link of this stage with intermolecular interaction.

No degradation was observed in all samples even at elevated temperatures. This indicates the absence of unstable interval length between chemical links. At the same time one can think that only few intervals of small size exist in the samples because degree of tension is relatively high.

Considering different kinds of deformation one can see that parameters of process subject to different conditions enable to evaluate free energy of system under tension. The specific functions describing free energy link to transformation from equilibrium distribution (as in Gibbs case) [15] of initial sample to equilibrium distribution of the sample after work made by tensile force.

Deformation of crosslinked sample results in various rearrangements at molecular and supermolecular levels. This probably enables estimation of physico-mechanical and other parameters, which link to particular stages of deformation, and dependence of these parameters on system structure and external conditions (temperature, stress).

Moreover, one of a way of current approach is a possibility of find particular thermodynamic stationary stages and the dependence of their parameters on external conditions. The consequence could be understanding of conditions of appearance of these stages and finding of a dependence of properties of particular sample on its pre-history.

REFERENCES

[1] A.L. Rabinovich. *Introduction in mechanics of reinforced polymers*. Moscow, Nauka, 1970

[2] A.D. Bernatskii. *Effect of structure of rigid polymer networks on parameters of general Maxwell's equation*. Moscow, Nauka, 1981

[3] E.F. Vainstein, E.S. Solodysheva, *Volume on Technology. Ser. "Articles produced from composites materials"*, 1995, 1, 12

[4] E.F. Vainstein, E.S. Solodysheva, *Volume on Technology. Ser. "Articles produced from composites materials"*, 1995, 2-3, 21

[5] T.I. Sogolova, G.A. Slonimsky. *Journal of Mendeleev Chemical Society*, 1961, 24, 6

[6] S.I. Zhurkov, T.I. Sepfirova. *Doklady of the USSR Academy of Sciences*, 1995, 12

[7] E.F. Vainstein, *Volume on Technology. Ser. "Articles produced from composites materials"*, 1996, 3-4, 43

[8] I. Prigozhin, P. Defei. *Chemical Thermodynamics*, Novosibirsk, Nauka, 1964

[9] I.N. Bronstein, K.A. Semendiaev. *Reference book on mathematics*, Moscow, Nauka, 1986

[10] B.M. Favorskii, A.A. Detlaf. *Reference book on physics*, Moscow, Nauka, 1964.

[11] A.E. Elsgoltz. *Differential equations and variation analysis*. Moscow, Nauka, 1969
[12] E.F. Vainstein, *Volume on Technology. Ser. "Articles produced from composites materials"*, 1995, 2, 30
[13] E.S. Solodysheva, E.F. Vainstein, *Articles produced from composites materials*, 2000, 2, 85
[14] E.F. Vainstein, *Articles produced from composites materials*, 1999, 3, 82
[15] L.D. Landau, E.M. Lifshitz, *Statistical physics*, Moscow, Nauka, 1964

Chapter 6

SYNTHESIS AND PHOTOPHYSICAL PROPERTIES OF HEXAZOCYCLANES-BIFLOUROPHORES AND MACROHETEROCYCLING COMPOUNDS, CONTAINING LANTHANIDES

S. A. Siling[a], S. V. Shamshin[a], A. V. Grachev[b], O. Yu Tsiganova[b], V. I. Yuzhakov[b], I. G. Abramov[c], A. V. Smirnov[c], S. A. Ivanovsky[c], A. G. Vitukhnovsky[d], A. S. Averjushkin[d] and Bui Chi Lap[d]

[a]Institute of Organo-Element Compounds Russian Academy of Sciences

[b] Moscow State University Physical faculty 119899 Moscow

[c] Yaroslavl Polytechnical Institute

[d] Physical Institute Russian Academy of Science

ABSTRACT

A method of the synthesis of macroheterocycles – hexazocyclanes-fluorophores, hexazocyclanes-bifluorophores, including hexazocyclanes -fluorophores and hexazocyclanes-bifluorophores containing rear-earth metals were developed. Photophysical properties of hexazocyclanes - fluorophores, and hexazocyclanes-bifluorophores were investigated. It was shown that in hexazocyclanes-bifluorophores non-radiation energy transfer takes place.

Keywords: absorbtion, fluorescence, hexazocyclanes-fluorophores, hexazocyclanes-bifluorophores, rear-earth metals, transfer of electron excitation energy.

AIMS AND BACKGROUND

The significant interest to problems of nanophysics is caused by perspective of creation of essentially new electronic devices having high speed (picoseconds), high reliability and very small sizes. The needs of nanophysic in many respects are supplied with significant successes of the chemists in creation of new materials having unique electronic properties on nanometres distances.

The development of methods of the synthesis of macroheterocycles (hexazocyclanes) - bifluorophores and hexazocyclanes, containing rare-earth metals is a real step in creation of materials for devices of nanoelectronics.

The research of luminescent characteristics and features of transfer of electronic excitation energy in conditions of spatial restrictions is an independent task. Research of influence of the ligand structure on optical properties nanoobject, containing rare-earth metals is obviously important.

It is known that lanthanides have unique luminescent properties: millisecond time of fluorescence and small width of spectrum of fluorescense[1] Lanthanides may be used for example for creation of amplifiers of signals in the lines of the optical communications, due to intensive fluorescence in near IR-area. In amplifiers of optical signals Eu frequently is applied because the length of wave of its radiation (1530 nm) practically coincides with widely used lengths of waves of signals in glass light conductor[2]. However bad solubility of cations of lanthanides in the appropriate inorganic mediums such as silicon glass results in clasterisation lanthanides, which limits the opportunities of aplification. It is result of cooperative processes of transfer of energy (processes of self-suppression) between lanthanides incorporated in clusters. It results in reduction of intensity of fluorescence of lanthanides. Large attention of the researchers is involved to increase of solubility of cations of lanthanides and carrying out of methods of production of homogeneous dispersion of lanthanides in glasses, and also engineering of reception of hybrid devices: optic light conductors and light-emitting diodes (LED).

However now solubility of lanthanides in real systems is limited by 0.1 wt.% of Ln^{3+}. The alternative to above-stated is use of transparent polymers instead of inorganic glasses. Organic ligands should be chemically attached to polymeric chain[3].

For achievement of isolation lanthanides in cluster i.e. amplification of luminescence, the introduction of lanthanides cations into macroheterocycles that separate cations from each other and consequently reduce processes of self-suppression is offered.

Therefore, the principal aim of our research is the synthesis of macroheterocycles-hexazocyclanes-fluorophores, hexazocyclanes-bifluorophores and hexazocyclanes-trifluorophores, containing rare-earth metals, study of their spectral-fluorescent characteristics and the opportunity of non-radiation transfer of energy of electronic excitation within macroheterocycles.

EXPERIMENTAL

Macroheterocycle on the Basis of 1-(3,4-dicyanophenyl)-benztriazole and 9,10-bis(4-aminophenyl) antracene (I)

1-(3,4-dicyanophenyl)-benztriazole (0.2450g, 0.001mol), 9,10-bis-(4-aminohenyl) antracene (0.3600g, 0.001 mol) and phenol (5g, 0.0467 mol) were placed in three-necked flask fitted with a condenser, thermometer and capillary for argon supply.

The reaction mixture was heated to 180°C, kept for 12h. to stoppage of ammonia formation, cooled to 50-60 ° C and poured into methanol. The beads were filtered off, washed with methanol, dried in vacuum over P_2O_5 for 12h. The mass of the beads 0,5370, yield-94%.

Found: C- 81.02; H-3.81; N- 14.02; calc. for $C_{80}H_{46}N_{12}$: C-81.75; H-3.94; N-14.3

Macroheterocycle on the Basis of 1-(3,4-dicyanophenyl)-benztriazole and rodamine (II)

1-(3,4-dicyanophenyl)-benztriazole (0.2450 g, 0.001 mol), rodamine (0.3489g, 0.001 mol) and phenol (5g, 0.0467 mol) were placed in three-necked flask fitted with a condenser, thermometer and capillary for argon supply. The reaction mixture was heated to 180°C, kept for 12 h. to stoppage of ammonia formation to the degree of the reaction completion -90-95%. Then the reaction mixture cooled and poured into methanol. The beads were filtered off, washed with methanol, dried in vacuum over P205 for 12 h. The mass of the beads 0,525, yield - 92%

Found: C- 70.41; H-3.85; N- 19.21; Cl-6.02; calc. for $C_{68}H_{46}N_{12}O_8$: C-70.83; H-3.50; N-14.57;C1-6.12

Macroheterocycle on the Basis of 1-(3,4-dicyanophenyl)-benztriazole and safranine T (III)

1-(3,4-dicyanophenyl)-benztriazole (0.2450 g, 0.001 mol), safranin T (0.3509g, 0.01mol) and phenol (5g, 0.0467 mol) were placed in three-necked flask fitted with a condenser, thermometer and capillary for argon supply. The reaction mixture was heated to 180 °C and kept to stoppage of ammonia formation. Then the reaction mixture cooled and poured into methanol. The beads were filtered off, washed with methanol, dried in vacuum over P_2O_5 for 12 h. The mass of the beads 0,5279, yield - 90%.

Found: C- 70.41; H-3.85; N- 19.21; Cl-6.02; calc. for $C_{68}H_{46}N_{16}Cl_2$: C-70.52; H-4.00; N-19.35; C1-6.12

Macroheterocycle on the Basis of o-dinitrile of phthalic acid and 9,10-bis(4-Amino-phenyl) antrace(IV).

o-dinitrile of phthalic acid (0.1280 g, 0.001 mol), 9,10-bis-(4-aminohenyl)antracene (0.3600g, 0.001 mol) and phenol (5g, 0.0467 mol) were placed in three-necked flask fitted with a condenser, thermometer and capillary for argon supply. The reaction mixture was heated to 180°C, kept for 12h. to stoppage of ammonia formation, cooled to 50-60°C and poured into methanol. The beads were filtered off, washed with methanol, dried in vacuum over P_2O_5 for 12h. The mass of the beads 0,5456, yield - 90%.

Found: C-86.11; H-4.17; N-8.61; calc. for $C_{68}H_{40}N_6$: C-86.79.; H-4.28; N-8.93

Macroheterocycle on the Basis of o-dinitrile of phthalic acid and unsubstituted rodamin (V)

o-dinitrile of phthalic acid (0.1280 g, 0.0024 mol), rodamine (0.8352g, 0.0024 mol) and phenol (5g, 0.0467 mol) were placed in three-necked flask fitted with a condenser, thermometer and capillary for argon supply. The reaction mixture was heated to 180°C, kept for 12h. to stoppage of ammonia formation, cooled to 50-60 ° C and poured into methanol. The beads were filtered off, washed with methanol, dried in vacuum over $P_2 0_5$ for 12 h. The mass of the beads 0,3450, yield - 92%

Found: C-72.81; H-3.40; N-8.92; calc. for $C_{56}H_{34}N_6 0_8$: C-73.19; H-3.73; N-9.14; 0-13.93

Macroheterocycle on the Basis of o-dinitrile of phthalic acid and 9,9-bis (4-aminophenyl) phluorene (VI).

o-dinitrile of phthalic acid (0.64 g, 0.005 mol), 9,9-bis (4-aminophenyl) phluorene (1,74g, 0.005 mol) and phenol (lOg, 0.0934 mol) were placed in three-necked flask fitted with a condenser, thermometer and capillary for argon supply. The reaction mixture was heated to 180°C, kept for 12h. to stoppage of ammonia formation, cooled to 50-60 ° C and poured into 100 cm^3 of the mixture methanol : water = l : l vol. The beads were filtered off, washed with water, reprecipitated from acetone to water, dried in vacuum over $P_2 0_5$ for 12 h. The mass of the beads 1,79g, yield - 80,54 %

Found: C- 86.02; H-4.35; N- 8.92; calc. for $C_{66}H_{42}N_6$: C-86.25; H-4.61; N-9.14

Eu-containing macroheterocycle on the Basis of phthalodinitrile and 9,9-bis(4-aminophenyl) phluorene (VII)

Macroheterocycle on the basis of phthalodinitrile and 9,9-bis(4-aminophenyl)phluorene (1,840g, 0.002 mol), europmniacetate (0.38 g., 0.00099 mol), N-methylpyrrolidone (10 cm) and tertiary butyl(2 cm^3) were placed in three-necked flask fitted with a condenser, thermometer and capillary for argon supply. The reaction mixture was heated to 160°C, kept for 6h. to stoppage of ammonia formation, cooled to 50-60 ° C and poured into water. The beads were filtered off, washed with water, dried in vacuum over P_2O_5 for 12 h. The mass of the beads 0,389g., yield - 70%.

Found: C-79.52; H-4.05; N-8.32; calc. for $C_{32}H_{22}N_{12}Eu$: C-79.75; H-4.15; N-8.44; Eu-7.63

Sm-containing macroheterocycle on the Basis of phthalodinitrile and 9,10-bis(4-aminophenyl) antracene (VIII)

Phthalodinitrile (0.128 g, 0.001 mol), 9,10-bis(4-aminophenyl) antracene (0.3600g, 0.001 mol), samarium chloride (0.0912 g., 0.00025 mol) and phenol (5g, 0.0467mol) were placed in three-necked flask fitted with a condenser, thermometer and capillary for argon supply. The reaction mixture was heated to 180°C, kept for 12h. to stoppage of ammonia formation, cooled to 50-60 ° C and poured into methanol. The beads were filtered off, washed with methanol, dried in vacuum over *P^Os* for 12h. The mass of the beads 0,235g, yield - 73%.

Found: C- 79.50; H-5.11; N- 7.56; calc. for $C_{144}H_{114}N_{12}Sm$: C-79.95; H-5.30; N-7.76; Sm-6.94

The absorption spectra of solutions of all investigated compounds in dimethyl-formamide were recorded at room temperature on a Specord M-40 instrument, spectra of fluorescence and excitation of fluorescence - on a Jobin Yvon 3CS instrument, ensuring automatic correction of spectra.

Results and Discussion

Earlier we developed a method of synthesis of azomethynes and polyazomethynes-bifluorophores[4,5]. It was shown, that in these substances non-radiating transfer of energy of electronic excitation is observed. The following stage of our work was the research of a possibility of non-radiation energy transfer in macroheterocycle. Macroheterocycles – bifluorophores were unknown earlier. As an object of the research macroheterocycles-hexazocyclanes[6] were chosen. They are synthesized by condensation of phthalodinitriles and diamines.

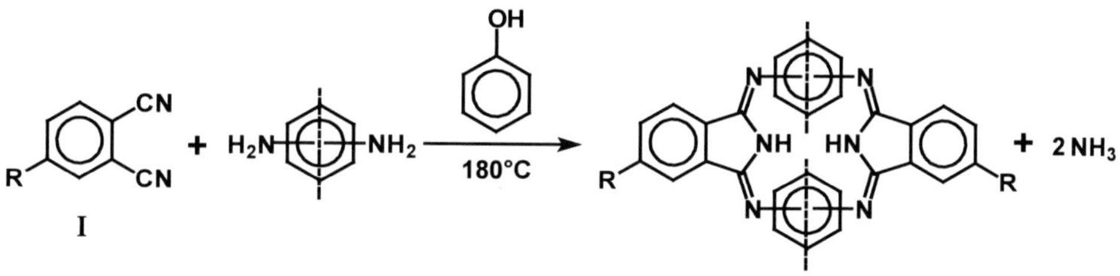

I

For synthesis of hexazocyclanes-bifluorophores phthalodinitriles-fluorophores, where one atom of hydrogen is substituted by heterocycle and phthalodinitriles-fluorophores with condensed heterocycles were chosen as donor of an energy. All of these phthalodinitriles were synthesized in Yaroslavl Politechnical University for the first time.

As a preliminary photophysical properties of solutions of phthalodinitriles-fluorophores in dimethylformamide were investigated. . Their spectral characteristics are shown in the table 1.

From the table 1 one can see , that 1 - (3,4-dicyanodiphenyl) benzimidazole has the widest spectrum of fluorescence, and also the most long wavelength of fluorescence. This substance was chosen as the donor of energy at synthesis of hexazocyclanes-bifluorophores.

Table 1 Photophysical Properties of o-phthalodinitriles-fluorophores

o-phthalodinitriles	Maximum of absorption, nm	Maximum of fluorescence, nm
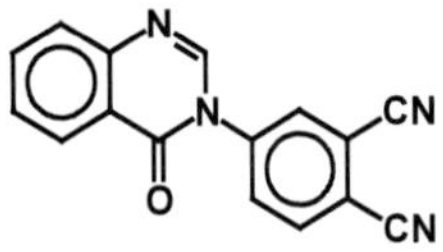	269, 280, 300, 315	445 500
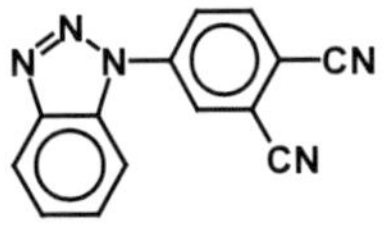	268, 320	432 500 550
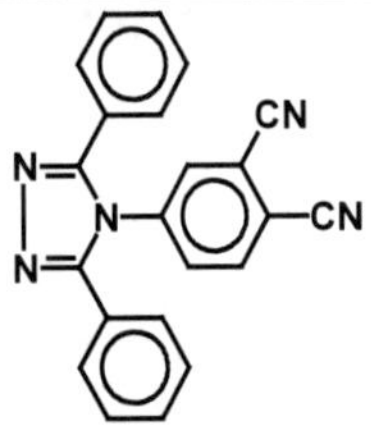	265, 297, 312	458
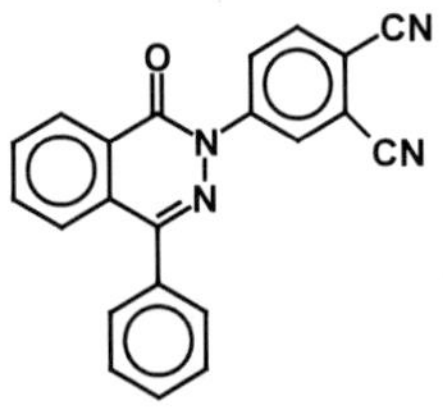	265, 305, 332	432
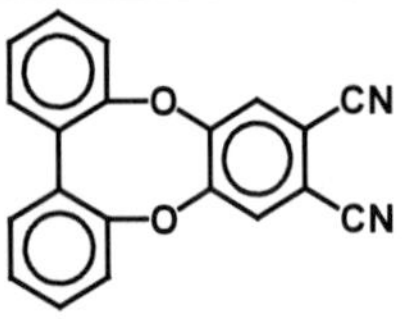	287, 299	360, 413, 434
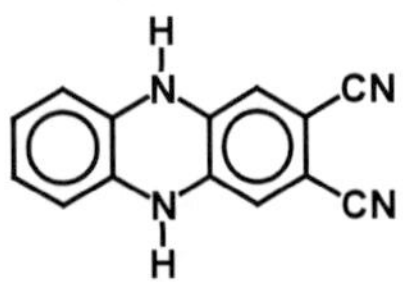	270, 300, 312	409
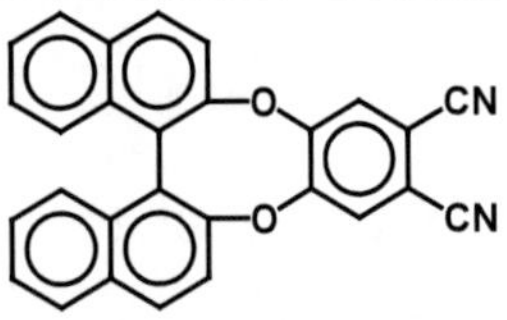	295	407

As an acceptor of energy the following diamines-fluorophores were used:

Table 2 Photophysical Properties of Diamines-Fluorophores

The structural formula of the diamine	Maximum of absorption, nm	Maximum of fluorescence, nm
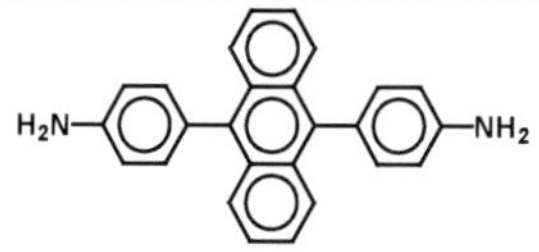	360, 380, 400	510
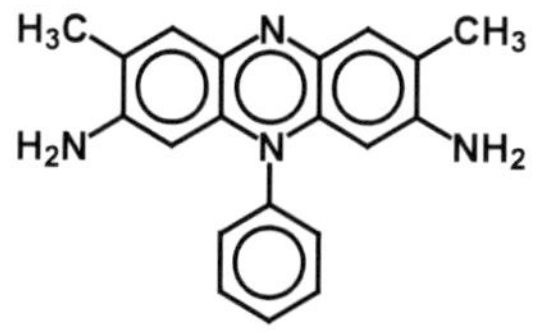	505,538	575
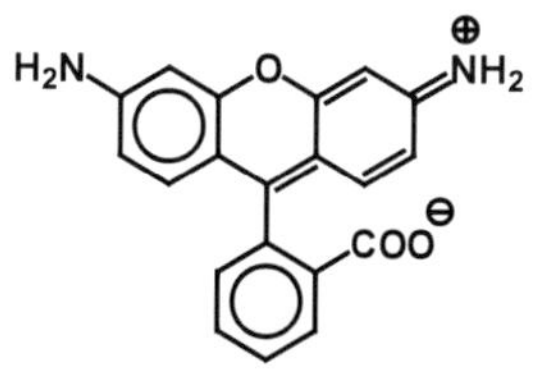	514	540, 560

The donor and acceptor of an energy were selected so that the spectrum of a radiation of the donor of an energy was as much as possible overlapped by the spectrum of absorption of the acceptor.

The synthesis of of hexazocyclanes-bifluorophores was carried out in phenol at temperature of its boiling. The reaction was controlled by the quantity of ammonia that is formed in the process of the reaction The reaction was considered as finished when the formation of ammonia was finished at a degree of the reaction completion - 90-95%.

In total 3 hexazocyclanes-bifluorophores (**I, II, III**) and 2 hexazocyclanes- fluorophores (**IV, V**) were synthesized.

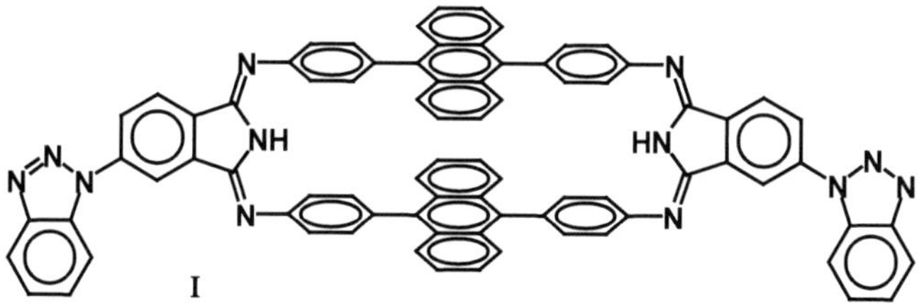

I

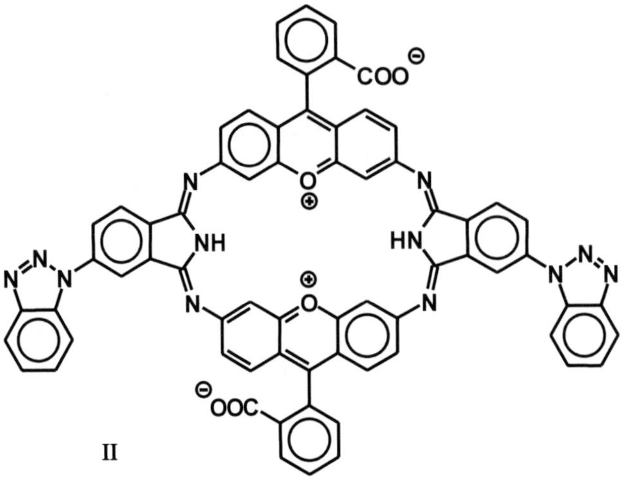

II

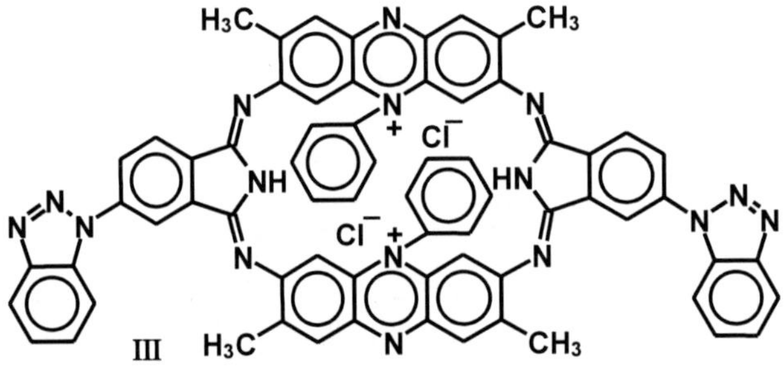

III

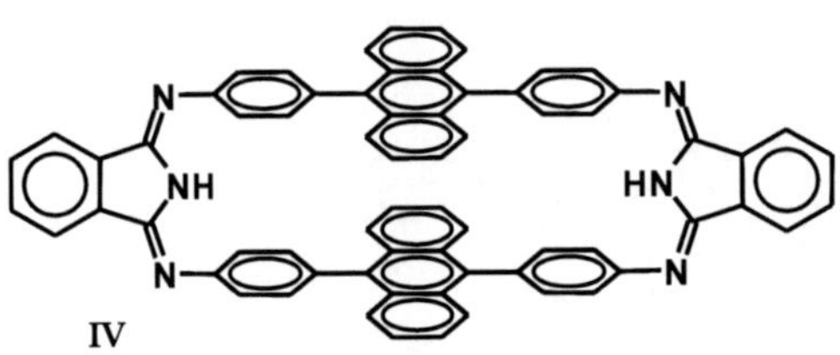

IV

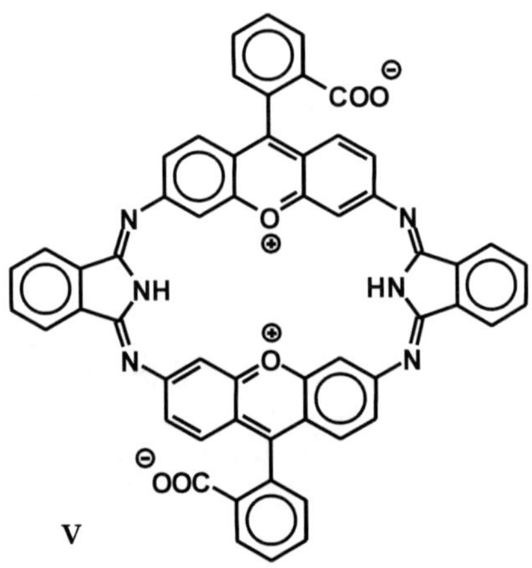

V

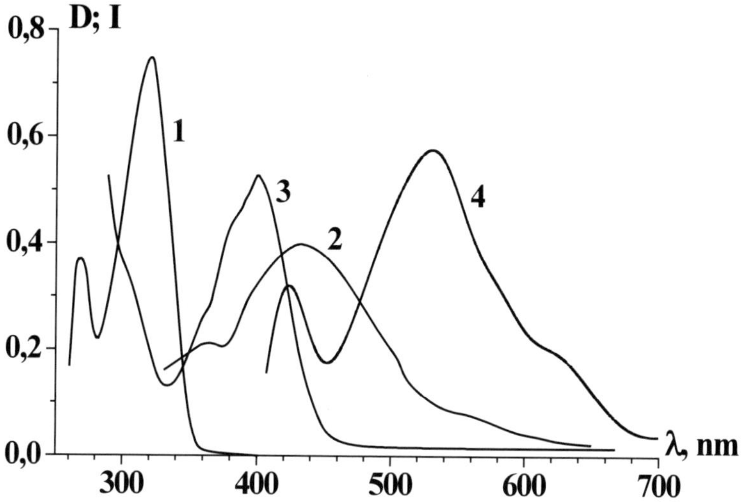

Fig. 1a Spectra of an absorption (1, 3) and fluorescence (2,4) of 1 - (3,4-dicyanophenyl) benztriazole (1and 2) and 9,10-bis - (p-aminophenyl) antracene (3 and 4).

In Fig. 1a the spectra of absorption (1 and 3) and fluorescence (2 and 4) of the solutions of 1 - (3,4-dicyanophenyl) benztriazole (1и 2) and 9,10-bis - (p-aminophenyl) antracene (3 and 4) in dimethylformamide are given. From Fig. 1 one can see the partial overlap of fluorescence field of phthalodinitryl and absorption field of 9,10-bis(p-aminophenyl) antracene.

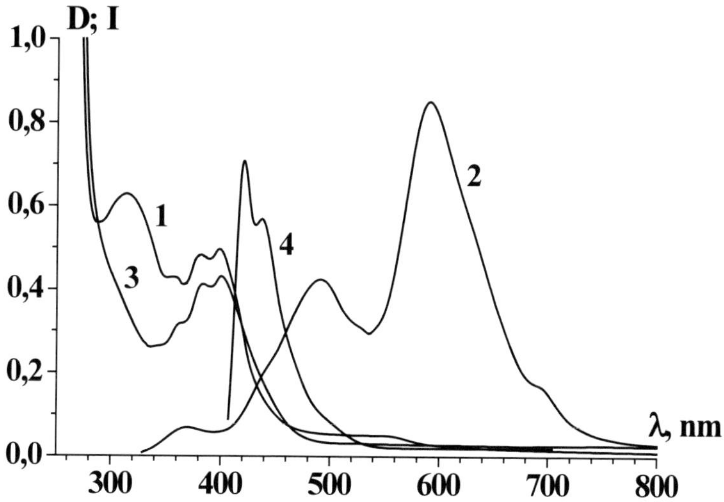

Fig. 1b Spectra of absorption (1,3) and fluorescence (2,4) of of hexazocyclanes **I** (1 and 2) and **IV** (3 and 4)

In Fig. 1b the spectra of absorption (1,3) and fluorescence (2,4) of solutions of hexazocyclanes **I** (1 and 2) and **IV** (3 and 4) in dimethylformamde are given. The spectra of fluorescence of macroheterocycles were received by excitation at a band of the excitation of donor energy fragment - 1 - (3,4-dicyanophenyl benztriazole) (λ = 312 nm). It is shown that there are bands of donor and acceptor of energy in spectrum of absorbtion of macroheterocycle. In the spectrum of fluorescence of macroheterocycle **I** 2 bands λ = 493 nm and λ = 590 nm and inflextion in the field of 700 nm are observed. The first band evidently belongs to isoindoline fragment of macroheterocycle, and the second band (the most intensive) belongs to bis(phenylantracene) fragment. The band (λ = 590nm) testifies non-radiating energy transfer from benztriazole to bis(phenylantracene) fragment of hexazocyclane **I**. Rather large batochromic displacement of maxima of fluorescece (approximately 70 nm) can be explained by hindered rotation of diphenylantracene fragment which is included in structure of macroheterocycle and additional energy obtained at its non-radiating transfer from benztriazole fragment to bis(phenylantracene) fragment. A new band of fluorescence in the field of 700 nm is explained, apparently, by the formation of a fragment with conjugated π-system.

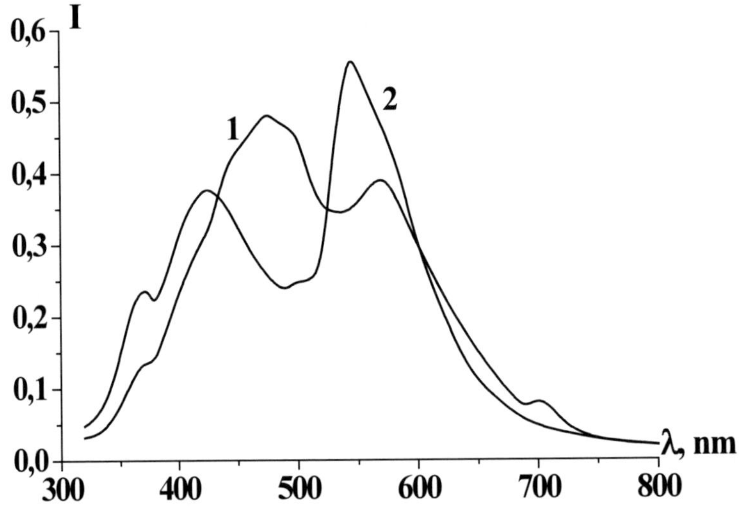

Fig. 2 Spectra of fluorescence of hexazocyclane II (1) and hexazocyclane V(2)

In Fig. 2 the spectra of fluorescence of hexazocyclane **II** (1) and the spectra of fluorescence of hexazocyclane **V** (2) received by excitation in a band of the absorption of benzthreeazole fragment (λ= 312 nm) are given. From Fig. 2 one can see , that in a spectra of fluorescence of hexazocyclane **V** (2) the band λ=540 nm, inflexion in the field of λ=560 nm, typical for unreplaced rodamine and band of fluorescence of isoindoline fragment-λ=430 nm are observed. In a spectrum of fluorescence of hexazocyclane **II** (1) a wide band in the field of λ=468-480 nm, which can be referred to fluorescence of substituted isoindoline fragment is observed. In a spectra also batochromic displacement of fluorescence band of rodamine fragment from λ=540 nm to λ=567 nm and the new band - λ= 698 nm, that may be explained by the formation of the system of conjugated links, are observed.

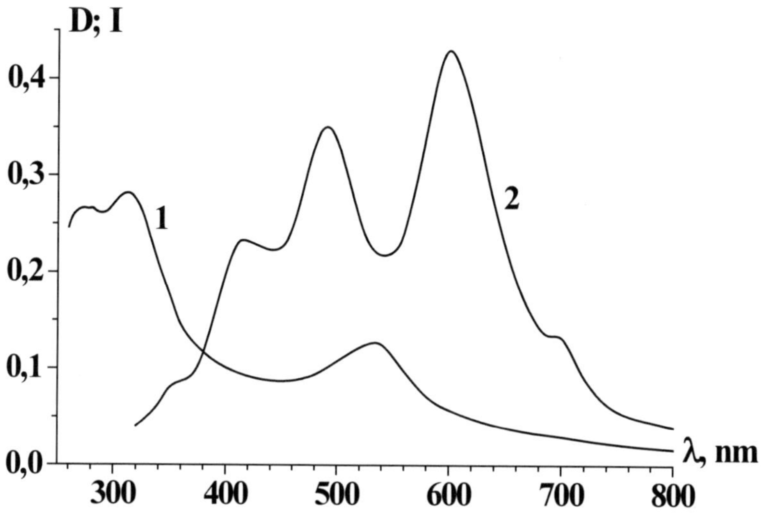

Fig. 3 The spectra of absorbtion (1) and fluorescence (2) of hexazocyclane **I**

In Fig. 3 the spectra of absorption (1) and fluorescence (2) of the solutions of hexazocyclane **I** in DMF are given. From Fig. 3 one can see, that at excitation in the and of an absorption of o-dinitrile – donor of energy (λ=312 nm) in a spectrum of fluorescence of hexazocyclane the band λ=596 nm is observed. It is very important that, in spite of the difference in the structure of macroheterocycles in the spectrum of fluorescence of hexazocyclane **I** as well as in the spectra of fluorescence of hexazocyclanes **III** and **II**, the band 698 nm is observed.

So for the first time macroheterocycles-bifluorophores were synthesized. It was established, that for these substances the non-radiating energy transfer at macroheterocycle is observed.

The following stage of our researches was the development of the method of synthesis of hexazocyclanes, containing rare-earth metals. The salt of rare-earth metals, and also crystallophores that contain rare-earth metals are rather widely used as luminophores at cathode and laser excitation. At photo excitation the intensity of their fluorescence is low, because the excitation of fluorescence takes place by transfer of electronic excitation energy from triplet level of an organic part of a molecule on metal. It is known that the energy of triplet level as a rule converts in non-radiation kinetic energy of a molecule. For the first time we have offered new approach to the solution of this problem, namely, an energy transfer for excitation of fluorescence of rare-earth metals not from triplet but from singlet level of ligand. For this purpose ligand should contain fluorescent fragments. As such ligands hexazocyclanes-fluorophores were chosen.

At the first stage the method of introduction of rare-earth metal in macroheterocycle, which does not contain fluorescent fragments was developed.

As an example the interaction of hexazocyclane **VI** on the base of phthalodinitrile and 9,9-bis - (p-aminominophenylfluorene) with europium acetate was developed.

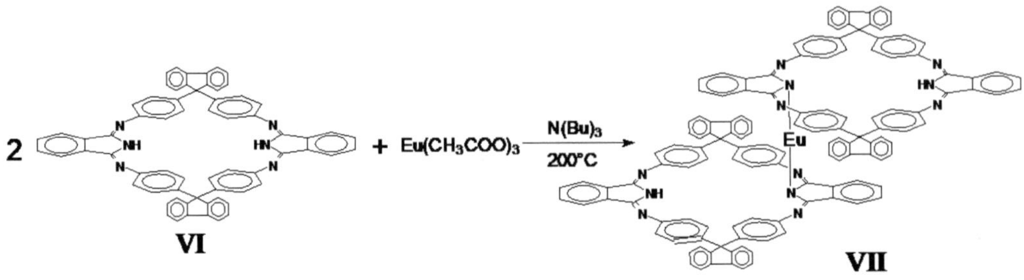

The reaction have carried out in tributylamine at 200°C in argon during 8 hours (to disappearance of initial substances according the date of thin layer chromotography). The obtained compound has a structure named "sendvich".

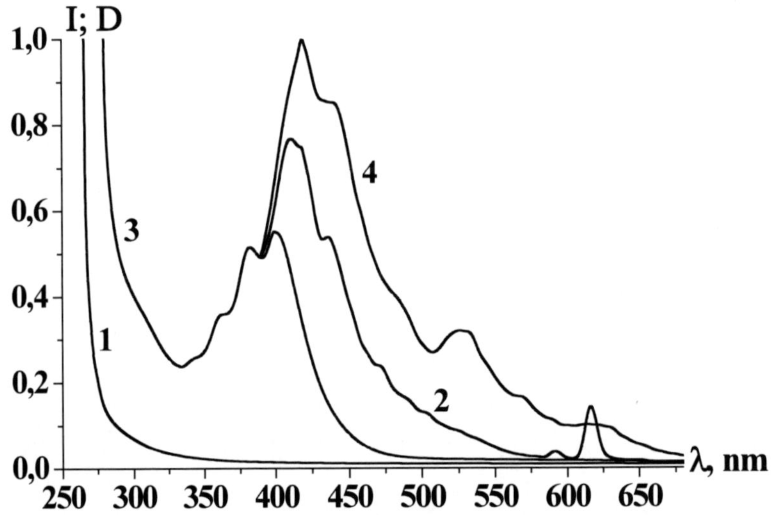

Fig. 4 The spectra of absorbtion (1,3) and fluorescence (2,4) of europium acetate and substance **VII**

In Fig. 4 the spectra of an absorption (1,3) and fluorescence (2,4) of solutions of europium acetate and substance **VII** in DMFA are given. From Fig. 4 one can see that in a spectrum of an absorption of europium acetate (curve 1) there are no bands in visible and near UV-field. It is cause of low intensity of fluorescence at photoexcitation. In the spectrum of absorption of substance **VII** (the curve 2) an intensive band f 300-450 nm is observed The spectrum of fluorescence are received by excitation at a band λ=365 nm

From Fig. 4 one can see also, that the square of a spectrum of fluorescence of hexazocyclane, containing Eu(III) (curves 4), is considerably more, than the square of spectrum of fluorescence of europium acetate (curves 3) in spite of the mass concentration of europium acetate in solution is by 10 times more, than the concentration of macroheterocycle, containing Eu(III). It means, that the intensity of fluorescence of hexazocyclane, containing Eu(III) exceeds the intensity of fluorescence of europium acetate more than 10 times in spite of the fact that macrocyclic ligand does not contain fluorescent fragments.

For energy transfer from singlet level of macrocyclic ligand to rare-earth metal macroheterocycle on the base phthalodinitrile and 9,10-bis (p-aminophenyl)antracene was used as a donor of an energy. Sm (III) was used and as an acceptor of energy. Metalcontaining macroheterocycle **VIII** synthesized by condensation of phthalodinitrile, 9,10-bis (p-aminophenylantracene) and samarium chloride in phenol at 180°C, in an atmosphere of an argon within 12 hours up to discontinuance of isolation of ammonia.

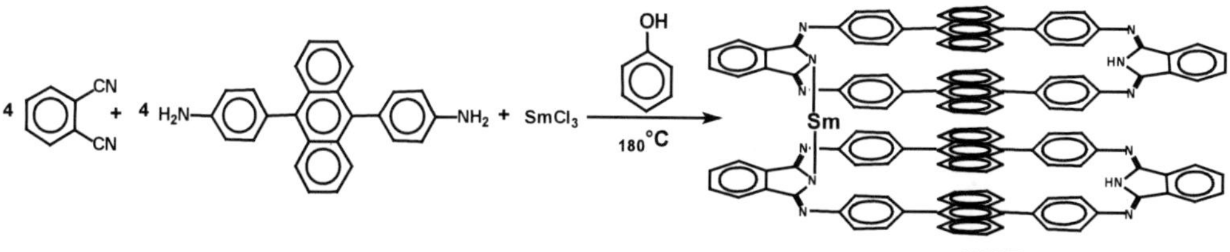

VIII

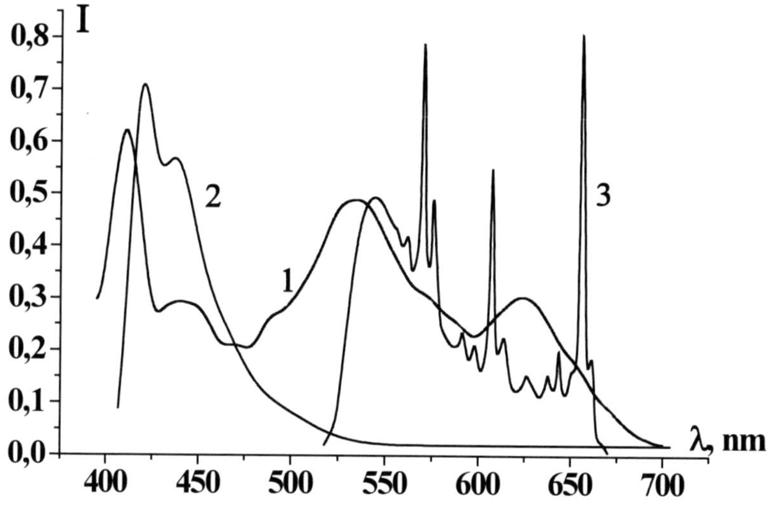

Fig. 5 Spectra of fluorescence of substance (**VII**) (1), macroheterocycle **IV** (2) solutions in DMFA and the crystal, containing ZnS, Cu and Sm (3).

In Fig. 5 the spectra of fluorescence of solutions of substance **VIII**(1), macroheterocycle **IV** (2) in DMFA and crystal, containing ZnS:Cu, Sm) (3) are given. From Fig. 5 one can see, that in the spectrum of fluorescence **VI** the intensive band in the field λ=400-475 nm is observed. In a spectrum of crystal the bands in the field (λ= 520-575 nm) and (λ= 600-670 nm) are observed. In the spectrum of fluorescence of **X**, received by excitation at a band of the absorption of bis (p-aminophenyl)antracene (λ=365 nm) 4 intensive bands are observed. The bands in the field of λ=400-450 nm concern to fluorescence of macrocycle-ligand. The bands in the field (λ=500-600 nm and λ=600-700 nm) are the bands of fluorescence of Sm. It unequivocally testifies non-radiating energy transfer from macrocyclic ligand on rare-earth metal.

For further increasing of quantum yield of fluorescence lanthanides the method of synthesis of hyperbranched macroheterocyclic polymers containing lanthanides was developed. These polymers were not described earlier.

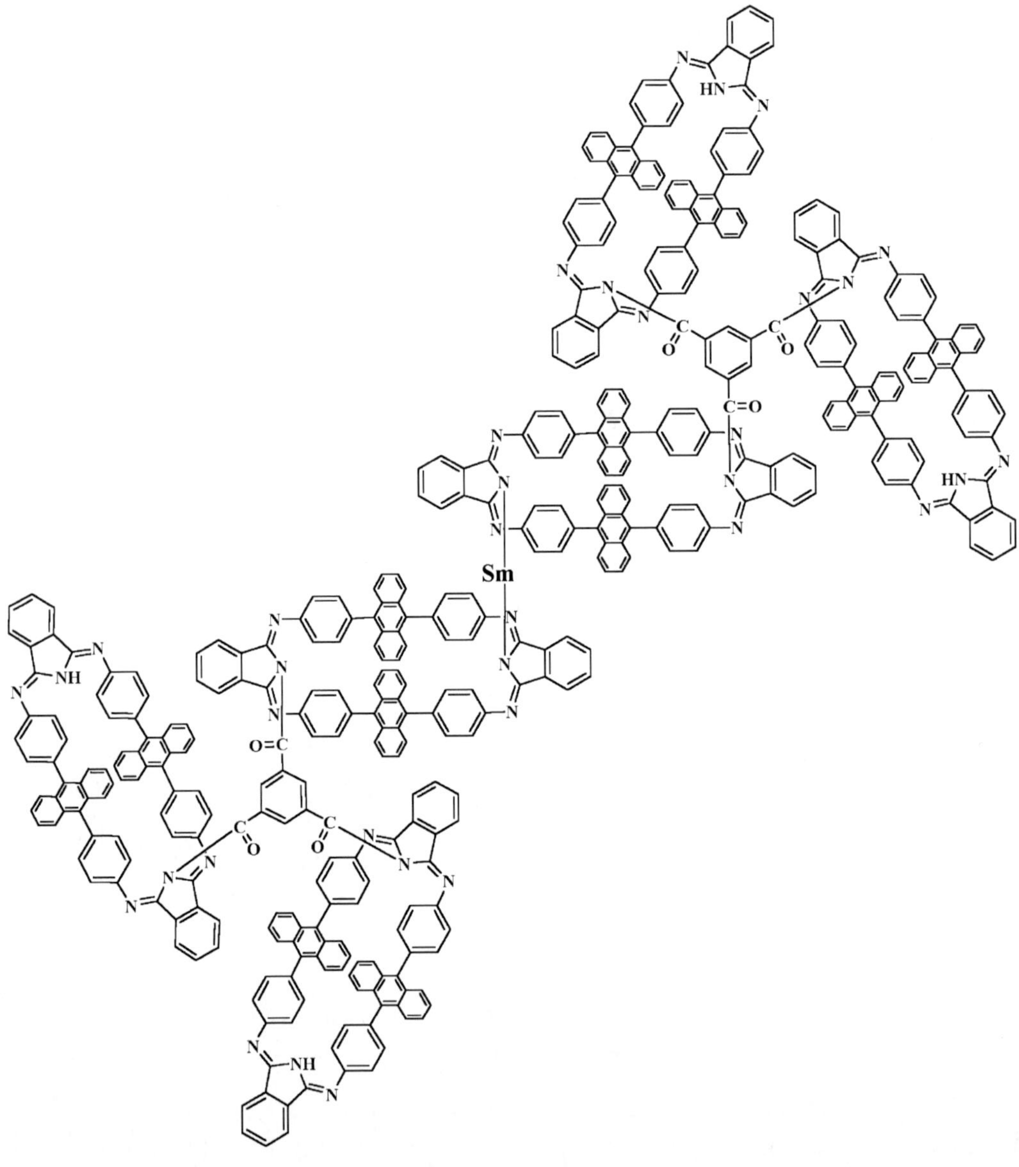

These polymers are photoantenna, where the energy of electron excitation is passed from fluorescent macrohetecocycles on rare-earth metal in this case on samarium. It is very important the fact that in this case the quantum yield of fluorescence of Sm, included in photoantenna exceeds 90%.

For increasing of the quantum yield of fluorescence of rare –earth metal the new type of hyper- branched macroheterocycling polymers were developed. In this case as monomers for the synthesis so named sandwich systems are used.

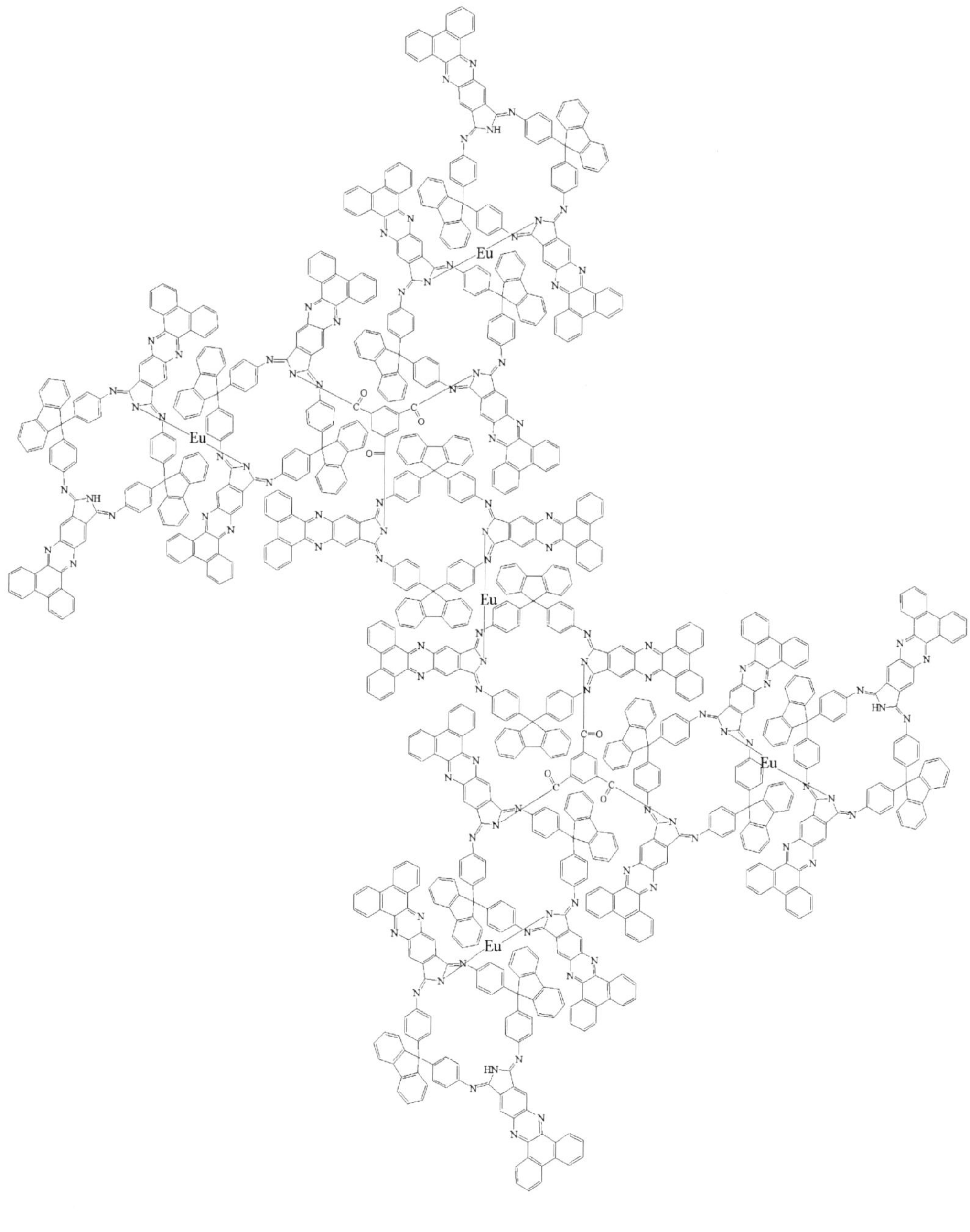

In this case the quantum yield of fluorescence of lanthanide is near 100 %

CONCLUSION

Thus, for the first time methods of synthesis of hexazocyclanes-bifluorophores are developed and their photophysical properties are investigated. It was established that in hexazocyclanes-bifluorophores the non-radiatinal transfer of energy of electron excitation takes place. For the first time hexazocyclanes-bifluorophores that contain rare-earth metals are synthesized. It was established, that the using even non-fluorescent macroheterocyclic ligands ensures photoexcitation of rare-earth metals. The intensity of fluorescence of rare-earth metals is sharply increased at a replacement of aliphatic ligand by macro-heterocyclic ligands . For the first time hexazocyclanes-bifluorophores in which the rare-earth metals are used as an acceptor of an energy of electronic excitation are synthesized. It was established, that in hexazocyclanes-bifluorophores, containing rare-earth metals the non-radiational transfer of an energy of electronic excitation from fluorescent fragment of macroheterocycle to rare-earth metals takes place.

REFERENCES

[1] Siling S.A., Shamshin S.V., Ronova I.A., et.al : Synthesis and photophysical ptoperties of azomethynes-bifluorophores, *Oxidation Communication* **22** (4), 495-508 (1999).

[2] Siling S.A., Shamshin S.V., Ronova I.A., et.al : Poly(Shif'c bases) – bofluorophores: synthesis, photophusical properties, non-radiational transfer of energy of electron excitation, *International Journal of Polymer Materials* **00**(1), 1-16 (2000).

[3] Siling S.A., Vinigradova S.V.: *Polycondensation of o-dinitriles of aromatic polycarboxylic acids and diamines – a new field of polymer chemistry*, Soviet Scientific Reviews (Section B, V.4) Harwood Academic Publishers (1992)

Chapter 7

COPOLYMERS WITH MACROHETEROCYCLES IN CHAINS COPOLYHEXAZOCYCLANES

S. A. Siting, S. V. Vinogradova, and N. M. Belomoina
A.N. Nesmeyanov Institute of Organo-Elemental Compounds,
Russian Academy of Sciences, Moscow, Russia

INTRODUCTION

One of the leading directions in polymer chemistry is synthesis and investigation of polyhetero-arylenes cyclo-chain construction. Among them polymers with macro-hetero-cycles in chains are especially interesting. The polycondensation of tetranitriles of tetracarbonic acids and aromatic diamines is unusual approach to synthesize macro-hetero-cyclic polymers. As the result of the polycondensation, polymers with macro-hetero-cycles in chains-polyhexazocyclanes (PHC)-are formed [1,2].

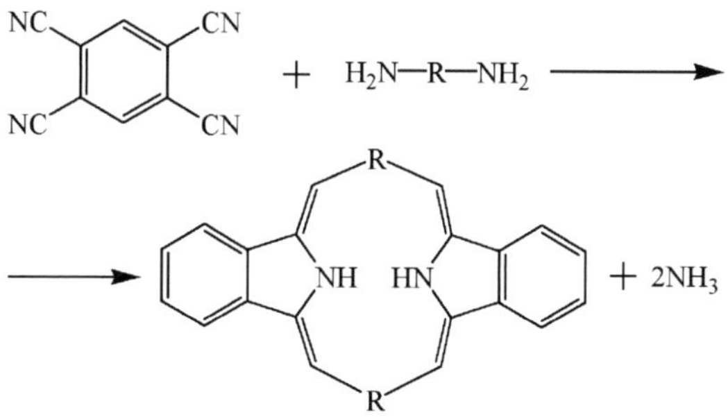

This method of synthesis allows to change structure and properties of PHC in wide ranges.

It was shown before [3] that the diamine structure affects significantly the polymer solubility, softening temperature, beginning of mass loss temperature. For example, PHC based on benzidine does not soften till temperature of intensive decomposition and dissolves only in 143804. Incorporation of oxygen bridge between phenyl nuclei makes polymer soluble in organic solvents and sharply reduces the softening temperature. At substitution of hydrogen atoms in phenyl nuclei by fluorine atoms, the range of solvents in which polymers dissolve becomes wider. Significant increase of polymer solubility can be achieved by incorporation of perfluorodimethylmethane group between phenyl nuclei of diamine.

This work devoted to investigation of PHC modification by changing the structure of the nitrite component. The basic direction of the modification was chosen synthesis of copolymers of regular structure- co-polyhexazocyelanes- i.e polymers in which macro-hetero-cycles alternate with chains of other polymers. This would give a goal-minded direction of PHC properties change.

The realization of such approach can be achieved by use tetranitriles of tetracarbonic acids containing fragments of different hetero-chain and hetero-cyclic polymers as one of the monomers. This would provide formation of co-polymers with high state of regularity and allow to hope to predict their properties with high probability.

Results and Discussion

Polyaryleneoxidehexazocyclanes

To obtain macro-hetero-cyclic polymers which would combine high heat stability and ability to process into articles in 'soft' conditions we sought to use as one of the monomers, tetranitriles of tetracarbonic acids containing ether groups- bis(dicyanophenoxy)arylenes(BCFA):

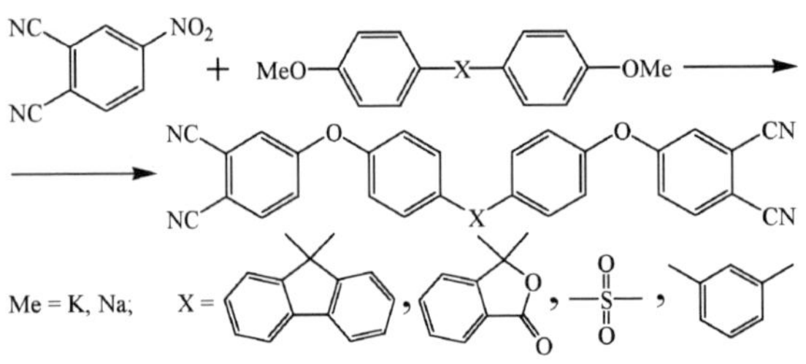

The PHC with maximal quantity of macro-hetero-cycles is formed by conducting reaction in phenol [1]. However, at polycondensation of BCFA with diamines in phenol, polymers precipitated as a gel at reaction completion of 25-60%. It was possible to consider that in this case the side reaction took place - interaction of BCFA with phenol on phenoxidic groups. This assumption was confirmed by investigation of reaction products of reaction of bis(3,4-dicyanophenoxy)sulfon with m-chlorophenol. In the mass-spectrum of the mixture ion peaks with m/z 250, 383, and 128 were found what gave a basis to suggest the following scheme of reaction:

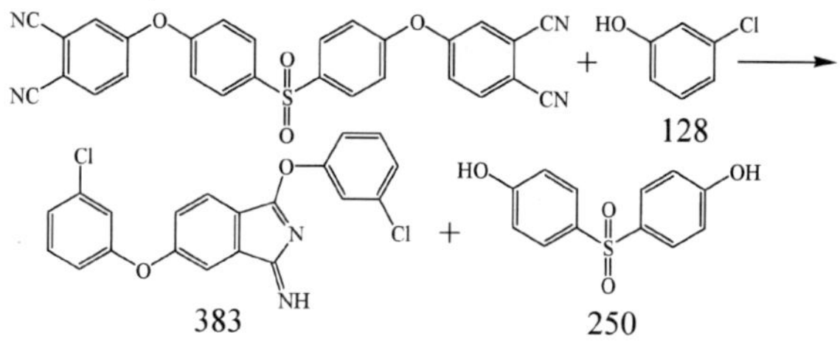

128 383 250

Because of this, the reason of gel formation at the reaction between BCFA and diamine could be interaction of formed bisphenol with nitrile end groups.

To avoid the gel formation, the products of interaction between BCFA with alcohol in presence of alkali metal alcoholates - bis(l-methoxy-3-iminoisoindol)arylenes- were used for synthesis:

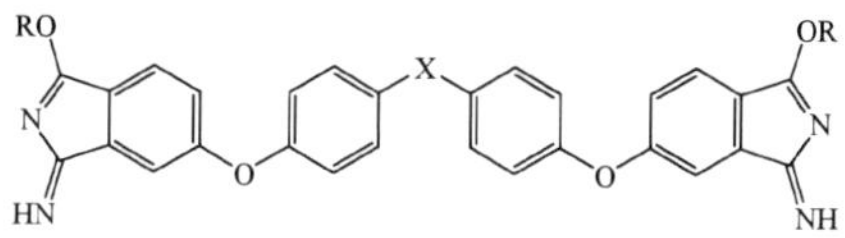

This allowed to conduct polycondensation of BCFA with diamines in methylpirrolidone (MP). As the result we obtained with quantitative yield macro-hetero-cyclic polymers - polyarylenoxide-hexazocyclanes PAOHC [4]:

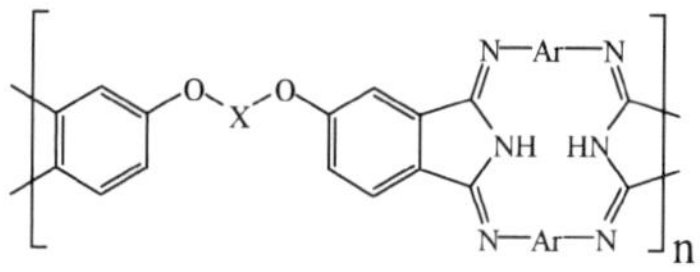

Some of the polymers properties are given in Table 1 .From the data in Table 1 it is seen that the polymers in deviation from PHC [1] soluble in wide range of solvents. The incorporation in nitrile component of cardovic (side cyclic groups) substituents provides positive influence on solubility as in case of diamines. At the same time, replacement of side phthalidic groups by fluorene groups leads to solubility increase and widening of range of solvents in which the polymers dissolve. The same groups provide increased polymer heat resistance. However, it should be noted that the effect is noticeable only at incorporation of

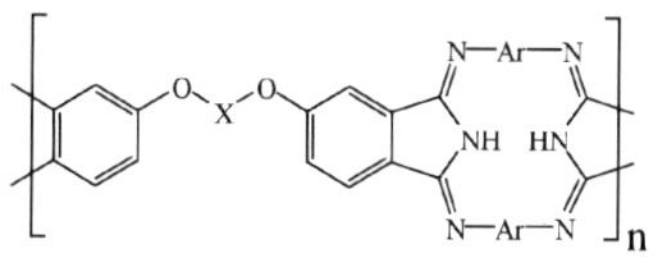

Table 1. Some Properties of Polyarylenoxidehexazocycles

X	Ar	Solvents in which polymer dissolves	Deformation starting temperature*, °C
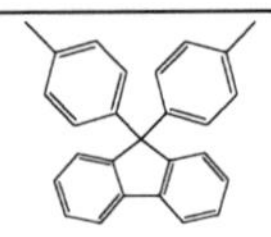	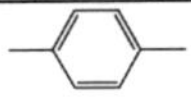	N-MP DMFA^ DMAA^^ Dioxane	250
-"-		-"-	310
-"-	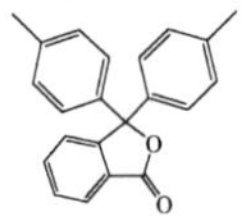	-"-	270
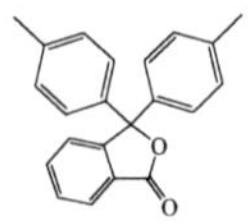	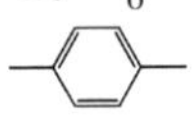	N-MP DMFA DMSO^^^ Dioxane	240
-"-	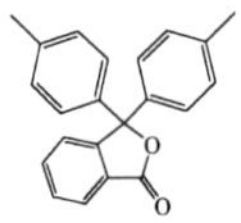	-"-	280
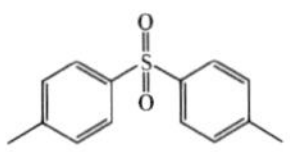	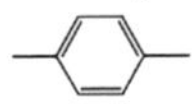	DMSO N-MP DMFA	130**
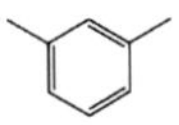	-"-	-"-	160**
-"-	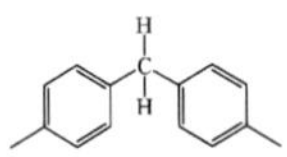	-"-	190**
-"-	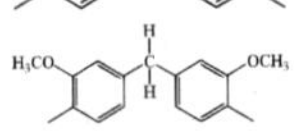	-"-	190**
-"-		DMSO N-MP DMFA Dioxane	220**

*At P = 2.45 kPa.. ^MFA-dimethylformamide

**At P = 0.78 Mpa. ^DMAA-dimethylacetamide ^^^DMSO- dimethylsulfoxide

cardovic groups in the nitrile component of PAOHC. The PAOHC have high thermostability. The temperature of beginning of mass loss in air is 400 -420°C based on the TGA data.

Polyester- and Polyamidohexazocyclanes

The necessity to provide polymers with macro-hetero-cycles in chains a high level of adhesion properties and ability to form films could be achieved by incorporation in the nitrile component of ester and

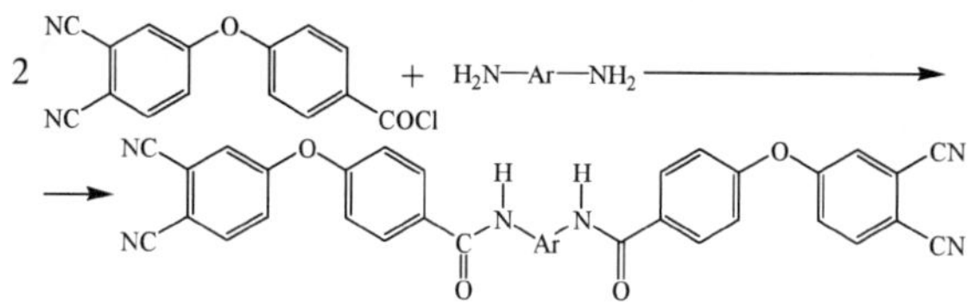

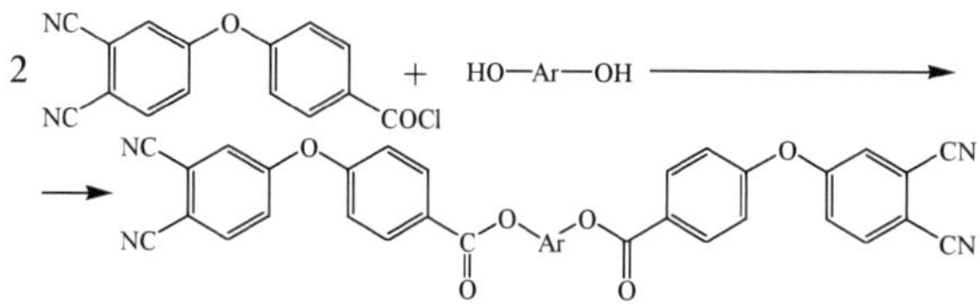

amide groups the cohesion energy of which is relatively high [5]. Existence of ester and amide groups in tetranitriles of tetracarbonic acids also eliminated the possibility to use phenols as a reaction media at their polycondensation with diamines. The low solubility of tetranitriles of such structure in alcohols did not allow to obtain compounds on their basis with end alcoxyiminoisoindol groups which could be usable as original products. Therefore, the possibility to use various solvents for synthesis of polymers directly from tetranitriles of tetracarbonic acids was investigated. The obtained results showed that polycondensation of tetranitriles of tetracarbonic acids containing amide and ester groups with diamines can be realized in MP in presence of Na alcoholate. Using this way, polyesterhexazocyclanes (PEHC) and polyamidohexazocyclanes (PAHC) were obtained.

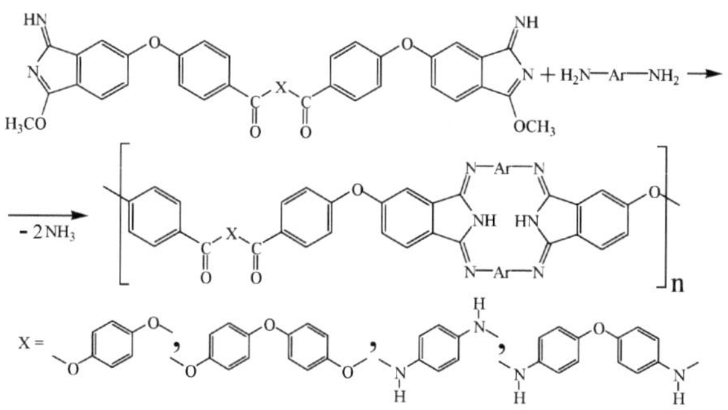

The deficiency of this method of synthesis is a low state of reaction completion (less than 60%) and as result of this - low (till 0.1 dl/g) intrinsic viscosity of 0.5% solution in MP.

The polyester- and polyamido-hexazocyclanes are obtained with high yield (0.94-0.95). They dissolved well in organic solvents, have high thermo-stability -the temperature at which the mass loss is 5% is 400 - 440 °C (See Table 2)

Effect of amide and ester groups is seen in that that the PEHC and PAHC form strong film on glass and metal due to probably increase of cohesion energy compared with PHC. However, the films prepared by pouring solutions of PEHC and PAHC in MP on solid surface have low strength properties. This is related probably both with low molecular mass of polymers and with relatively low content of ester and amide groups in their elementary chains.

Table 2. Properties of Polyester- and Polyamido-hexazocyclanes of Common Formula:

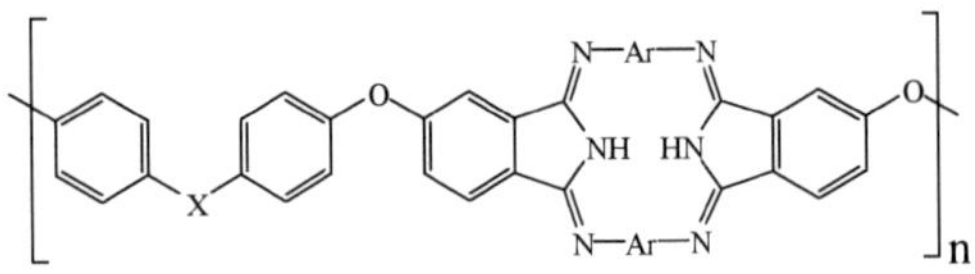

X	Ar	η_{sp}*, m^3/kg	T_{st}^, °C	T_5', °C
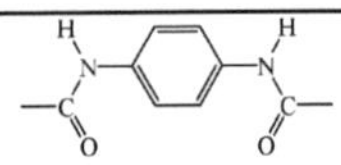	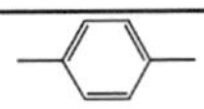	0.08**	290	440
-"-	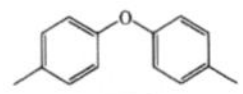	0.09	275	420
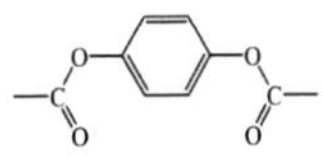	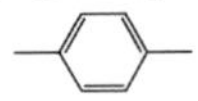	0.07	270	420
-"-	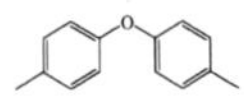	0.06	190	400
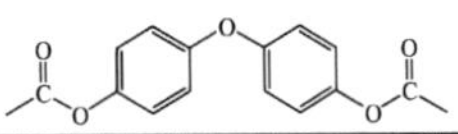	-"-	0.09	160	420

* 0.5% solution in N-MP.

^ Temperature of softening start taken from the thermo-mechanical curve.

' Temperature at which mass loss in air is 5% based on the TGA data. The heating rate was 4.5 °C/min.

** η_{sp} of 0.25% solution in formic acid - 0.33 m^3/kg

Polyphenylquinoxalinhexazocyclones

Polyphenylquinoxalines during many years attract attention of investigators [6,7] due to good solubility of polymers, high thermal and deformational properties, ability of processing into articles by usual technological methods. This was the basis for incorporation in the polyhexazocyclanes of phenylquinoxalinic fragments. To realize the modification of PHC, the bis(dicyanophenoxy)heteroarylenes with phenylquinoxalinic cycles were used as the original compounds for synthesis.

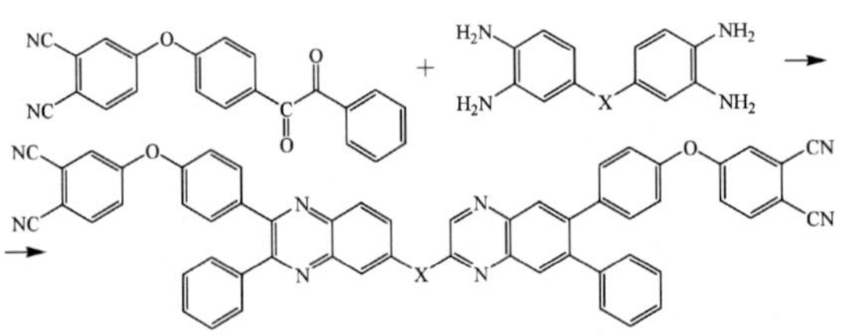

By polycondensation oftetranitriles oftetracarbonic acids with phenylquinoxalinic cycles and aromatic diamines a new class of macro-hetero-cyclic polymers – polyphenylquinoxalinhexazocyclanes (PPQHC)-was obtained.

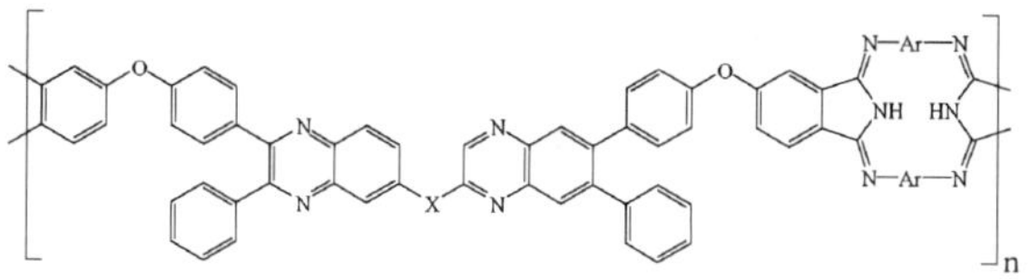

As it is seen from Table 3, PPQHC are obtained with a good yield and relatively low mass. Their solubility is noticeably worse compared with known polyphenylquinoxalines what is related probably with existence of macro-hetero-cycles in their polymer chains. The polymers have a high thermo-stability. The temperature at which mass loss is 5% in air based on TGA is in the range 430-470 °C.

The influence of phenylquinoxalinic fragments was wound to be the most effective in decreasing of co-polymer heat stability. Because of this the PPQHC process into articles very well by heat pressure molding. The PPQHC are able to form structure at temperatures above 270 °C due to interaction of amine- and nitrile end groups. The heat treated articles made of PPQHC do not soften till the temperature of intensive decomposition what is a big advantage of these materials compared with the articles made of polyphenylquinoxalines. The PPQHC materials form strong coatings on glass and metal.

Table 3. Properties ofpolyphenylquinoxalinehexazocyclanes of common formula:

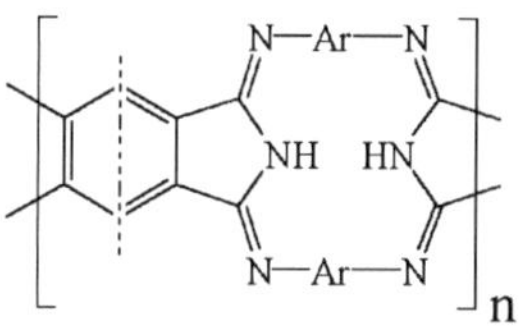

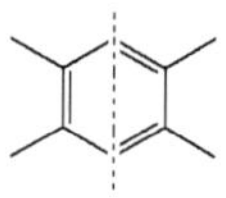	**Ar**	**Polymer yield, %**	**η_{sp}*, m^3/kg**	**Temp. of softening start^, °C**	**T_5**, °C**
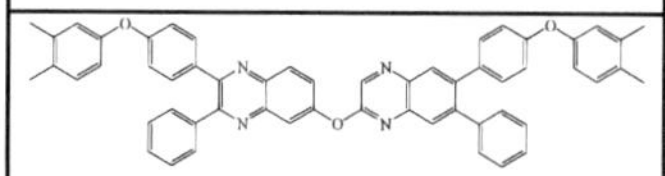	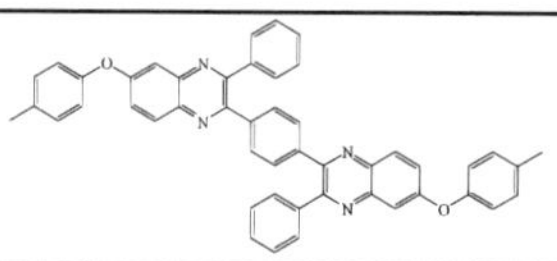	95	0.12	280	440
-"-	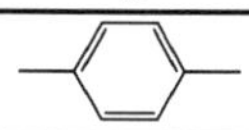	95	0.10	230	430
-"-	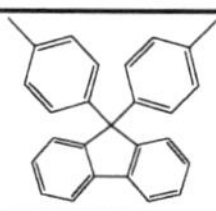	92	0.10	280	440
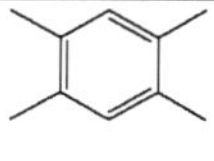	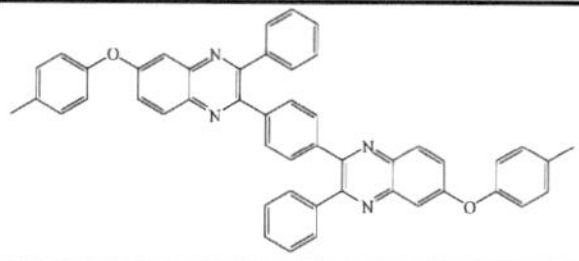	95	0.12	290	450
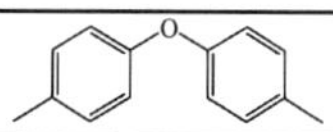	-"-	96	0.10	280	440

* 0.5% solution of polymer in HCOOH.

^ The temperature of softening start obtained from thermo-mechanical curve.

** Temperature at which mass loss in air is 5% based on the TGA data. The heating rate was 4.5 "C/min.

Polyquinazolonhexazocyclanes

Polyheteroarylenes obtained from bisbenzoxazinones and diamines -polyquinazolones- even in the row of ladder polymers show a high level of thermo-stability along with ability to process easily into articles [8,9]. Because of this, it was worthwhile to investigate the effect of incorporation of quinazolonic fragments into PHC on their properties. The tetranitriles of tetracarbonic acids containing benzoxazinonic fragments were obtained according to the following scheme [10]:

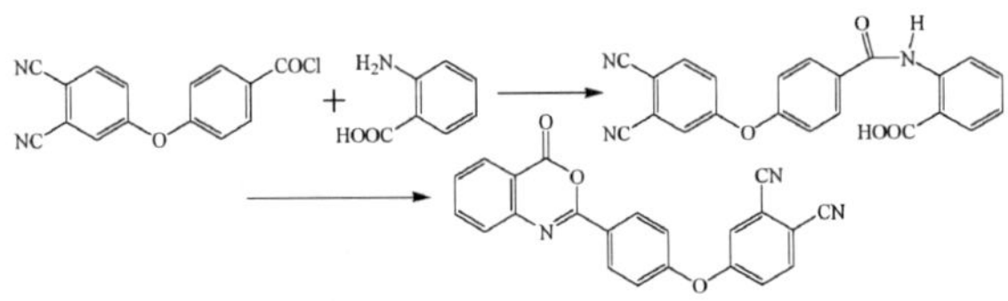

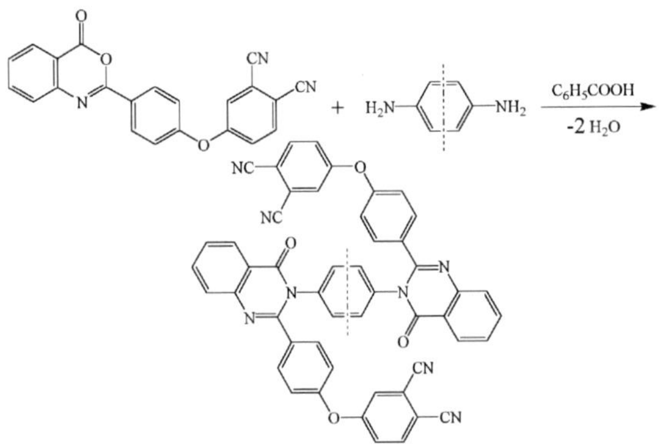

By polycondensation of tetranitriles with diamines the following polyquinazolonhexazocyclanes (PQHC) having the structure shown on the next scheme were obtained.

The PQHC are powders of dark color well dissolved in organic solvents. By pouring the polymer solutions on glass or metal, coatings can be obtained. By incorporation into PQHC a quinazolonic cycle, the thermo-stability of polymers can be enhanced. The temperature at which mass loss reaches 5% is 470-490 °C.

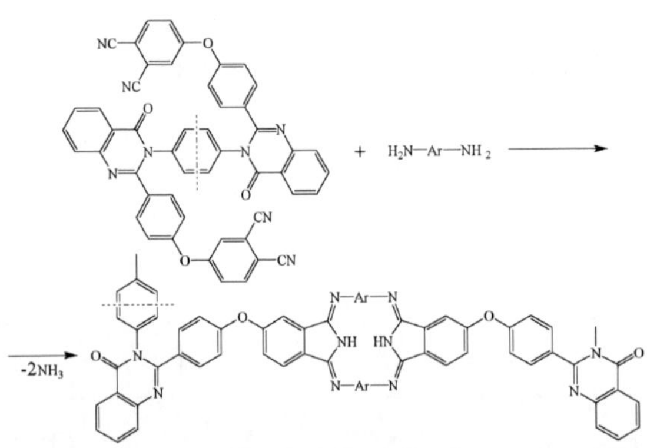

It should be noted that the processing temperatures of these polymers are lower compared with the ones of PHC (See Table 4).

Comparison of the results obtained at studying of properties of co-polyhexa acids with phenylquinoxalinic cycles and aromatic diamines a new class of macro-hetero-cyclic zocyclanes allowed to determine existence of clearly expressed dependence between the structure of the intermacrocyclic fragment and polymer properties. Thus, the most heat stability in the row of investigated polymers have the polyester- and polyamidohexazocyclanes which at P=0.8 kg/cm^2 do not soften till the temperature of intensive decomposition what is in deviation from other polymers. The same groups as well as phenylquioxalinic and quinazolonic cycles give polymers ability to form coatings on glass and metal. Incorporation of hetero-cycles in the intermacrocyclic fragment as it was shown on examples of PPQHC and PQHC leads to polymer themno-stability increase, especially in case of PQHC. These polymers have the most interesting set of properties because they combine good solubility and ability to form coatings with high thermo-stability and low processing temperatures to make articles by heat press molding.

Table 4. Properties of Polyquinazolonhexazocyclanes of Common Formula:

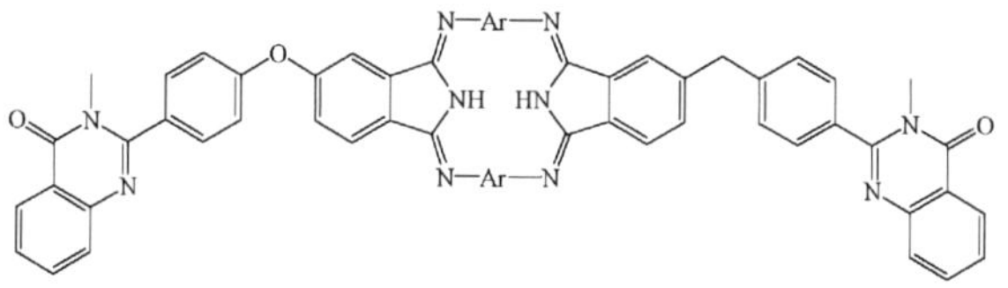

Ar	Polymer yield, %	η_{sp}*, m^3/kg	T_{soft}^, °C	T_5**, °C
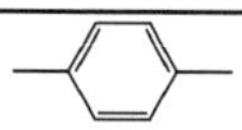	95	0.09	250	470
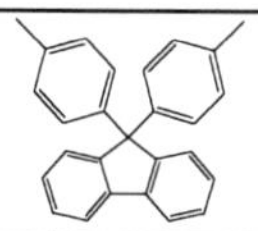	95	0.08	270	490

* 0.5% polymer solution in HCOOH

^ Temperature of softening start obtained from thermo-mechanical curve. Heating rate was 2 °C/min., P = 0.078 MPa.

** Temperature at which mass loss in air is 5% based on the TGA data. The heating rate was 5 °C/min.

Thus, incorporation of ester-, ether- amide-phenylquinoxaline-, and quinazolonic-intermacrocyclic fragments into main polymer chain leads to improvement of PHC properties and is a perspective method of polymer modification.

Experimental Part

Bis(l-methoxy-3-iminoisoindol)arylenes

A 0.003 mol solution ofNa methylate in 2 ml ofmethanol was added to 0.03 mol solution of bis(dicyanophenoxy)arylene in 20 ml of methanol and was kept for 12 hrs at room temperature. After that, the reaction mixture was poured into water, the precipitate was filtered off, washed with water till neutral reaction, dried in vacuum over H_2SO_4 for 24 hrs. The yield was 80%. The IR-spectrum showed the following bands: 2830 cm-' (-OCH_3) and 3400 cm-' (-NH). The band 2200 cm" (-C=N) was absent. ***Polyarylenoxidehexazocyclanes*** were prepared by using procedure published in [4].

Polyamidohexazocyclanes and Polyesterhexazocyclanes

Tetranitriles of tetracarbonic acids containing amide- or ester groups (0.001 mol), diamine (0.002 mol), MP (6 ml), and Na methylate (0.004 ml) were loaded into reactor. The reaction was conducted in a stream of argon at 200°C for 6 hrs till stoppage of ammonia deliberation. Then, the reaction mixture was poured into methanol, the precipitated polymer was filtered off, reprecipitated from DMFA into acetone, dried in vacuum at 60°C. The yield was 85-90%.

Polyphenylquinoxalinhexazocyclanes

Reactor was loaded with 0.8629g (0.001 mol) of bis-2-4'-(3,4-dicyanophenoxy)phenyl-3-phenylquinoxalino-6-oxide, diamine (0.002 mol), 0.153g of quinoline (0.0001 mol), 0.043g (0.0001 mol)ofLiCl and 10ml of phenol. The reaction was conducted in stream of argon at 175°C for 15-20 hrs. After that, the reaction mixture was cooled off to 60°C, poured into methanol, the polymer was filtered off, reprecipitated from chloroform into methanol, dried in vacuum at 60°C. The yield was 90%.

Polyquinazolonehexazocyclanes

Reactor was loaded with 0.002 mol of 2-4'[(3,4-dicyanophenoxy)phenyl] benzoxazinone, 0.001 mol of diamine, 0.002 mol ofbenzoie acid and 10 ml of phenol. Reaction was conducted at 100-110°C for about 8 hrs till disappearing in IR-spectrum of the reaction solution the band 1760 cm" corresponding to the C=0 group of benzoxazinone cycle. Then, 0.0002 mol ofLiCI and 0.002 mol of diamine were added, the temperature was risen to 175°C and the reaction continued till stoppage of ammonia deliberation. After completion of the reaction, the mixture was pored into methanol, the precipitated polymer was filtered off, repreeipitated from MP into methanol, dried in vacuum at 60-100°C. The yield was 85-99%.

REFERENCES

[1] Siting S.A., and S.V. Vinogradova 'Poly condensation of o-Dinitriles of Aromatic Polycarboxylic Acids and Diamines- a New Field of Polymer Chemistry'. Harwood Academic Publisher. *Sov.Sci.Rev., Sec.B*, v.76, Part 1.1992.

[2] Siling S.A., and S. V. Vinogradova *Uspechi Chimii*, v.63. No. 9, 1994, p. 810

[3] Siling S.A.. S.V Vinogradova, V.V. Korshak. and V.N. Solovjov *J. Polym. Sci.*, v.22, No.5, 1984,, p. 1043.

[4] Ustinov V.A.. I.I. Plakhtinskii, G.S. Mironov, V.A. Ryabukhina, V. V. Korshak, S. V. Vinogradova, and S.A. Siling, Sov. Patent 697502, *Bull. Invent.*, No.42, 1979. p.95.

[5] Kardashev D.A., "*Synthetic Glues*" (Russ.), M. Chimiya. 1976.

[6] Acutgu F., S. Kuze, K. Matsuok, K. Naruchi, and N, Miura//Macromol.Chem., *Rapid Commun.*, V.I 2, No.] 2. 1990,p.673.

[7] Fang T. *Macromolecules*, v.24, No.2. 1991, p.444

[8] GribkovaP.N., N.K. Vinogradova, Z.F. Saikina, Yu.E. Doroshenko, E.K.Skubin, and V.M. *Musykantov Vysokomol. Soed.*, A, v.31, No.5, 1989, p.1008.

[9] Shigery K.., M. Toshimoto, A. Torahiko, and A. Fukami *J. Polym. Sci., Polym. ChemEd.*, A, v.25, No.8, 1987,p.2241

[10] Kopeikin V.V., V.A. Ustinov, V.V. Sosnina, G.S. Mironov, I.I. Ponomarev, S.A. Siting, S.V. Vinogradova, and V.V. Korshak, Sov. Patent, 1,013,445 *Bull Invent.*, No.15. 1983, p.l 16.

Chapter 8

NEW GENERATION OF ZINC-AIR ENERGY SOURCES

S. A. Siting and L. A. Golubeva
A. N. Nesmeyanov Institute of Organo-Element Compounds (INEOS), RAN, Moscow, Russia
M. I. Bazanov and A. V. Petrov
*Ivanovo State Chemical Technology University, Ivanovo, Russia
S. K. Bychkovskii
**Nauchno-Proizvodsvennoye Predpriyatie (Company) "QUANT", Russia

A new type of metal-containing polymers has been developed in INEOS, Russian Academy of Sciences. It was found that the polymers can be successfully applied as a catalyst in the blend of active mass of carbon electrode for zinc-air energy sources (elements) – original and rechargeable batteries of general and special applications, and also for heating elements. The zinc-air elements are applied for energy supply to devices and different apparatus in which the ordinary manganese-zinc (MZ) or lithionylchloride (LC) energy sources are used. The size of zinc-air energy sources is the same as the size of the lithionylchloride ones.

The polymeric catalysts are very effective in composition of carbon cathodes for Li-thionylchloride (LTC) energy sources of special applications where the can replace silver and platinum.

The Figure below shows the discharge curves for non-catalyzed and catalyzed carbon cathodes of LTC energy sources.

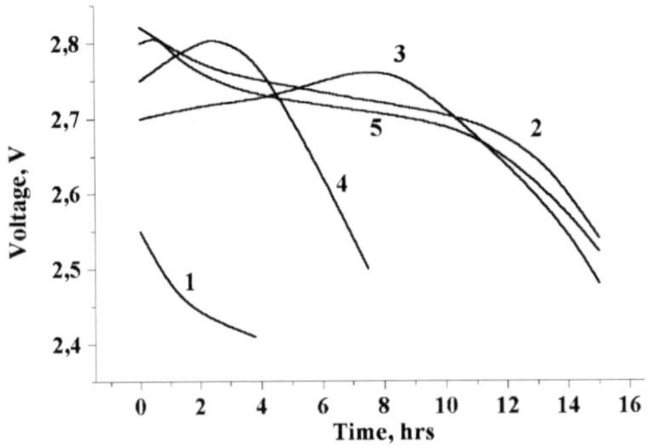

Figure 1. Discharge curves of Li-thiochloride energy sources: 1 – non-catalyzed carbon cathode; 2-5 – catalyzed carbon cathodes

It is seen from Fig. 1 that incorporation of polymeric catalysts into carbon cathode leads to discharge time increase from 3 to 15 hrs (compared with non-catalyzed cathode). This allows to increase the voltage from 0.10 V to 0.15 V, and the effective capacity of elements several times depending on the discharge current.

The catalyzed carbon cathodes can be used as full or partial replacement of MnO_2 cathode in the zinc-air energy sources with salt or alkali electrolyte (Table 1).

Table 1. Discharge characteristics of energy sources with alkali electrolyte and different cathode materials

Voltage, V	MnO_2-cathode	Catalyzed carbon cathode
Normal	1	1.1
Reduced discharged	to 0.8 A - l hr	to 0.9 V - 6 hrs
Electrical capacity (at current 0.5A)	450-1000 mA/hr	3000 mA/hr

From the data in Table 1, it is seen that the catalyzed carbon cathode allows to increase the effective capacity in 3 times, the discharge current -in 3.3 times, the working voltage - on 100 mV, and provide high (to 3-4 A/hr) electrical capacity of the air elements with alkali electrolyte at discharge current of 0.5A. This is six times higher than in ordinary energy sources. At the discharge current 0.5A, the discharge capacity of the commercial elements is 2.0-2.4 A·hrs, and the one of the elements based on the catalyzed carbon cathode is 3-4 A·hrs what is 1.5 times higher.

The discharge characteristics of energy sources with salt electrolytes are given in Table 2.

Table 2. Discharge characteristics of energy sources with salt electrolyte and different cathode materials

Discharge regime*	Discharge time, hrs	
	MnO_2 cathode	Catalyzed carbon cathode
On 4 hrs during 24hrs to voltage 0.9V	85	110-120
Continuous discharge to voltage 0.85V**	120	145-150

* Initial voltage 1.4 V, R = 20 Ohm.

** Resistance R = 10 Ohm

It is seen from Table 2 that at the continuous discharge regime, the elements with MnO_2 cathode show the reduced electrical capacity from 1.4V to 0.85V for 120 hrs. By using the catalyzed carbon cathode, the same electrical capacity decrease is achieved for 150 hrs. Thus, the duration of the last element is increased by 30%.

Based on the performed investigations, a zinc-air energy source (ZA) has been developed. In the element, carbon activated by polymeric catalyst was used as the positive electrode instead of manganese dioxide. The catalyzed carbon cathode allows to use air oxygen located beyond the energy source (element) to generate electricity what is significant advantage compared with the known elements. The high activity of the cathode allows to decrease the positive electrode size and increase the size (volume) of the negative electrode (anode). Due to the fact that the electrical capacity increases proportional to the anode material amount, the capacity of suggested ZA elements is higher compared with the one of the best alkali elements in 2-3 times. The results of comparative tests of size 316 elements which were conducted by different companies and the suggested element are given in Table 3.

Table 3. Comparative characteristics of suggested ZA element and the world best energy sources

Element type	Discharge duration, hrs		
	Continuous discharge, R_c=3.9 Ohm, final voltage 0.8V	1 hr per 24 hrs, R_c=3.9 Ohm, final voltage 0.8V	1 hr per 24 hrs, R_c=10 Ohm, final voltage 0.9V
'Samsung' (R6), salt	1.5	2.5	5.5
'Toshiba' (R6), salt	2.0	2.5	6.25
'Varta' (LR6), alkali	5.75	5.75	14.5
'Duracell' (LR6), alkali	-	7.0	19.0
'CTAPT (LR6), alkali	-	4.5	12.0
Suggested element	12-13	12-13	30-32

It follows from the data in Table 3 that the suggested zinc-air elements at indicated discharge regimes have duration superior in 2-3 times compared with the alkali elements, and in 6-8 times compared with the salt ones. The retail price of suggested elements is 67% of the best alkali elements produced by Duracell, Energizer; GP. The catalysts for ZA cathodes of the elements are produced based on the commercially available materials.

The production method is simple enough and does not require special equipment. The energy sources with carbon-catalyzed cathode are stable at storage. For example, after storage for 2 months, the energy sources with alkali electrolyte lost only 7% of the original voltage. The capacity was reduced on 14%. After storage for one year, the voltage decreased by 9%, the capacity - by 20%.

Chapter 9

THE NATURE OF ORGANOALUMINUM COMPOUND AND KINETIC HETEROGENEITY OF ACTIVE SITES IN LANTHANIDE-BASED DIENE POLYMERIZATION

N. N. Sigaeva, T. S. Usmanov, G. E. Zaikov and Yu. B. Monakov

Institute of Organic Chemistry Ufa Research Center
Russian Academy of Science
pr. Oktyabrya, 71, Ufa, 450054, Russia
Tel/Fax: [3472] 35-61-66

SYNOPSIS

The influence of organoaluminum compound on kinetic heterogeneity of active sites in lanthanide-based diene polymerization was investigated.
It was shown that heterogeneity of investigated catalytic systems has been displayed in the existence of four types of active centers. They are formed at the beginning of polymerization process and produce macromolecules with the lengths, which are definite for each type of active centers. The organoaluminum compound nature greatly influences on the kinetic activity of polymerization centers.
The method of analysis of curves' maximums of distribution on kinetic activity and the change of kinetic activity of every active centers' type was proposed. It permits to calculate the values of concentration for each type of active centers separately.

Key words: kinetic heterogeneity of catalytic system, active site, polymerization, butadiene, Tikhonov regularization method.

INTRODUCTION

The study of diene polymerization on lanthanide catalytic systems showed [1-4], that the nature of organoaluminum compound (OAC) essentially had influenced both on catalytic system activity and properties of the received polydienes. It was established [1,2], that the use of various OAC defined values of constants of transfer on OAC (k^{Al}_t) reaction rate and the values of propagation chain reaction rate constants (k_p) as well. The use of different OAC may influence on the microstructure of formed polydienes [3]. Molecular characteristics of forming polydienes significantly depend on the OAC content and nature. It was shown, that for the polybutadiene received on $NdCl_3 \cdot 3L$-OAC (L – tributylphosphate) catalytic system, the broad molecular weight distribution was typical. Form of the last one depends on synthesis conditions and OAC nature [4]. It was supposed, that the reason of the broad molecular weight distribution was the existence of several types of active centers [5,6].

In this connection, the influence of OAC nature on kinetic heterogeneity of active centers in lanthanide-based diene polymerization has been investigated.

EXPERIMENTAL PART

The butadiene polymerization was carried out at 25°C in the presence of the two-component lanthanide catalytic systems $NdCl_3 \cdot 3L$-OAC, where L was tributylphosphate. As organoaluminum compounds $Al(i\text{-}C_4H_9)_3$, $Al(C_2H_5)_3$, $Al(C_6H_{13})_3$, and $Al(C_8H_{17})_3$ were used. The complexes were prepared according to the technique described in [7]. A complex was formed by an *in situ* procedure, i.e. the solutions of OAC ($C_{Al}=3 \cdot 10^{-2}$ mol/l) and $NdCl_3 \cdot 3L$, were added sequentially into a toluene monomer (C_m=1,5 mol/l) solution at 25°C. The ratio Al/Nd was 30.

The 1,4-cis-units content was 90-95%, the content of 1,2-units was not less than 1%, and the 1,4-trans structures were the rest.

Molecular weights (M_w – weight-average, M_n – number-average) and the molecular weight distribution (MWD) of 1,4-cis polybutadiene were estimated by gel permeating chromatography (chloroform, 1 ml/min, 25°C). Four "Waters" columns packed with 10^6Å, 10^5Å, 10^4Å and 10^3Å μ-styragel were used. The system of columns was calibrated under the standards with narrow MWD ($M_w/M_n \leq 1,2$) using the universal dependence of Benoit [8] and equation connecting the molecular weight of polydienes with characteristic viscosity [9].

Chromatograms and the obtained values of molecular weights were corrected for instrumental spreading as described in [10].

The calculation of distribution function on kinetic heterogeneity of $NdCl_3 \cdot 3L$-OAC catalytic system was carried out by Tikhonov regularization method [11-14]. The minimum of Tikhonov functional was reached at regularization parameter $\alpha=9,4 \cdot 10^{-5}$ at quadratic average error of experiment $\delta^2=0,0001$. Method and procedure IUPR used in the work permit to modify error value in the choice of integral equation kernel. Since the kernel representating the Shultz-Flory function was chosen as a model, we assumed that error $h \rightarrow 0$.

RESULTS AND DISCUSSION

In a general view the broad polymers MWD may be represented by [15]:

$$u = PSX, \tag{1}$$

where u – polydispersity of polymer, is magnitude equal to the ratio of weight-average molecular weight to number-average molecular weight; P is measure of the correlation between polymerization reaction and diffusion limits; S is a factor depending on distribution of polymerization active centers ($S\to 2$ in the case of homogeneity of active sites); X is a function of particle size distribution.

As the butadiene polymerization process with lanthanide catalytic systems is homogeneous it is possible to assume that the P=1 and X=1. The kinetic regularities of diene polymerization on these catalysts confirm this supposition. The polymerization reaction is characterized by the first orders organomagnesium monomer and catalyst. The chain transfer reactions proceed under polymerization, but they don't result in the active centers deactivation. These facts testify that the MWD of polydienes, obtained on these systems, should be close to the most probable Flory distribution with M_w/M_n~2. However in most cases magnitude of polydispersity is significantly greater. Practically observed polydispersity of distributions on molecular masses for polydienes is significantly higher than that of theoretically expected, because active centers, differing in reaction ability are involved in polymerization process. There is some quasicontinuous distribution of active centers, each of them produces fractions distributed according to the law $\beta_i e^{-\beta_i M}$. Assuming that the law of catalytic centers distribution described by some function $\varphi(\beta)$, the MWD will be:

$$q_w(M) = \int_0^{\infty} \varphi(\beta)\, \beta^2 e^{-\beta M}\, d\beta \tag{2}$$

where β - the magnitude, is inverse the average molecular weight M_n, characterizing probability of a chain transfer and equals $\beta = \dfrac{1}{m_0} \cdot \dfrac{r_t}{r_p}$ (r_p – chain propagation rate, r_t – summarized rate of chain transfer and m_0 – molecular mass of monomer).

As it follows from expression (2), the values of $q_w(M)$ depend on an aspect of a cumulative distribution function on kinetic activity of catalytic system $\varphi(\beta)$. Suppose, $s = \ln \beta$ and $x = \ln M$, then it is possible to show, that $Z(s) = e^{2s}\varphi(e^s)$, $K(x, s) = \exp(x + s + \exp(x + s))$, $U_\delta(x) = q_w(M)$, and the equation (2) is reduced to a classical equation of the Fredholm of the first sort:

$$U_\delta(x) = \int_0^{\infty} Z(s)\, K(x, s)\, ds \tag{3}$$

Where $K(x, s)$ – kernel of an integral equation, and $Z(s) = e^{2s}\varphi(e^s)$. $Z(s)$ - unknown function, and the index δ in the left part of the equation specifies that the function, found from experiment, $U_\delta(x)$ – practically founded, is defined with an error δ.

The complexity of calculating $\varphi(\beta)$ from the equation (3) is in the fact that it has to do with the class of ill-posed problems [11-14]. So direct solution of such tasks because of an error of experimental data (the left part of the equation 3) may even reduce to absurd outcomes.

For the wide MWD (M_w/M_n>3) numerical solution of an integral equation (3) is carried out by the Tikhonov method of regularization by a selection of regularization parameter α, at which experimentally defined and calculated from the equation (2) functions $q_w(M)$ coincide with the given exactitude δ. Besides the obtained solutions $\varphi(\beta)$ should satisfy a condition of normalization:

$$\int_0^{\infty} \varphi(\beta)\, d\beta = 1 \tag{4}$$

Since β is a magnitude inverse to molecular weight, the solution of equation (3) in this work is represented as a plot of $\varphi(\beta)$ in the $\psi(\ln\beta)$-lnM coordinates, where $\psi(\ln\beta) = \beta\varphi(\beta)$.

The course of the polybutadiene molecular weights dependence on monomer conversion is analogous for all used catalytic systems (Figure 1). At the initial stage of polymerization M_w grows, when the conversion achieves 20-40% this growth ceases, and the further conversion growth leads either to the constant values of molecular masses or to the small reduce of M_w. Such molecular mass change is typical for the ion-coordinated diene polymerization processes and the invariable number of active centers, the occurrence of the chain transfer processes and monomer reducing during polymerization cause it.

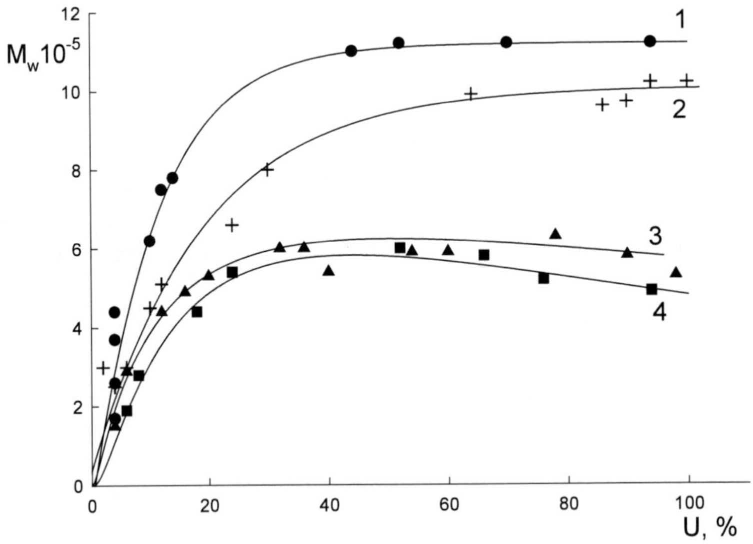

Fig. 1. The plots of M_w for polybutadiene vs. the monomer conversion under polymerization on $NdCl_3 \cdot 3L$-OAC catalytic system. OAC: 1 – $Al(i\text{-}C_4H_9)_3$, 2 – $Al(C_2H_5)_3$, 3 – $Al(C_8H_{17})_3$, 4 – $Al(C_6H_{13})_3$. Polymerization conditions: toluene, C_M=1,5, $C_{Al}=3 \cdot 10^{-2}$ mol/l, Al/Nd=30, T_{pm}=25°C.

The most high polybutadiene molecular masses are observed under the $Al(i\text{-}C_4H_9)_3$, then $Al(C_2H_5)_3$, $Al(C_8H_{17})_3$, and $Al(C_6H_{13})_3$ usage under equal conditions (Figure 1). As the conversion increases (Figure 2) molecular weight distribution curves are displaced into the higher molecular weight region, polydispersity of polymer increases by this. The growth of polydispersity during polymerization is observed for all investigated catalytic systems (Figures 2-5). However, the form of polybutadiene MWD curves is changed under the use of various OAC. As it follows from the data of Figures 2 and 3 MWD curves of polybutadiene, received with the use of $Al(i\text{-}C_4H_9)_3$ and $Al(C_6H_{13})_3$, are monomodal. Polymer synthesized on $NdCl_3{\cdot}3L$- $Al(C_2H_5)_3$ is characterized by bimodal distribution. MWD curves of polybutadiene received with the $Al(C_8H_{17})_3$ as OAC change their form under polymerization process. Broadband appears on the curves in the region of molecular masses' high values at high monomer conversions.

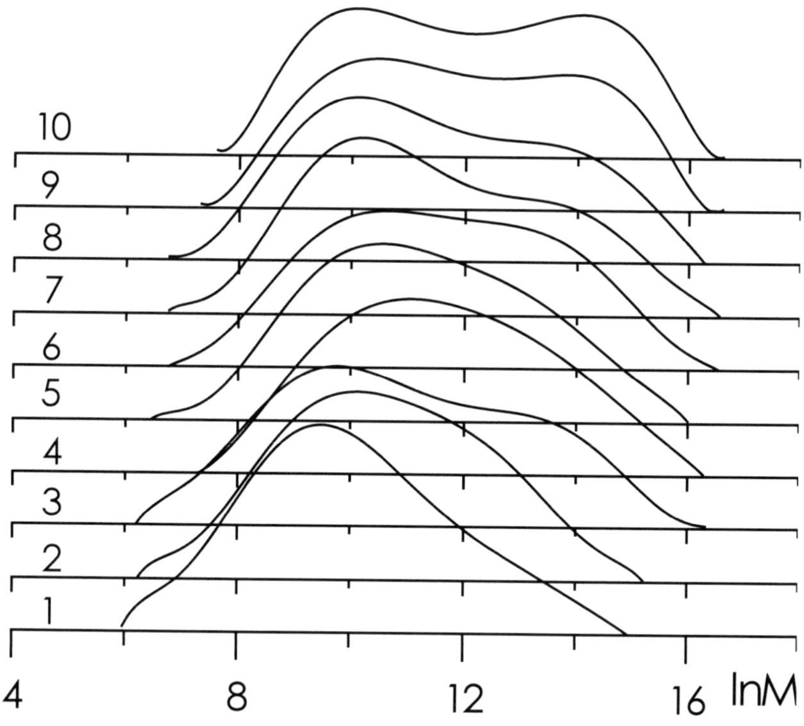

Fig. 2. Molecular weight distribution curves for polybutadiene, received on $NdCl_3{\cdot}3L$-$Al(i\text{-}C_4H_9)_3$ catalytic system. Polymerization conditions are given in Fig. 1. Monomer conversion (%):1 – 3, 61; 2 – 6,8; 3 – 11,1; 4 – 17,4; 5 – 25,6; 6 – 37,3; 7 – 55,6; 8 – 67,8; 9 – 87,9; 10 – 88,8.

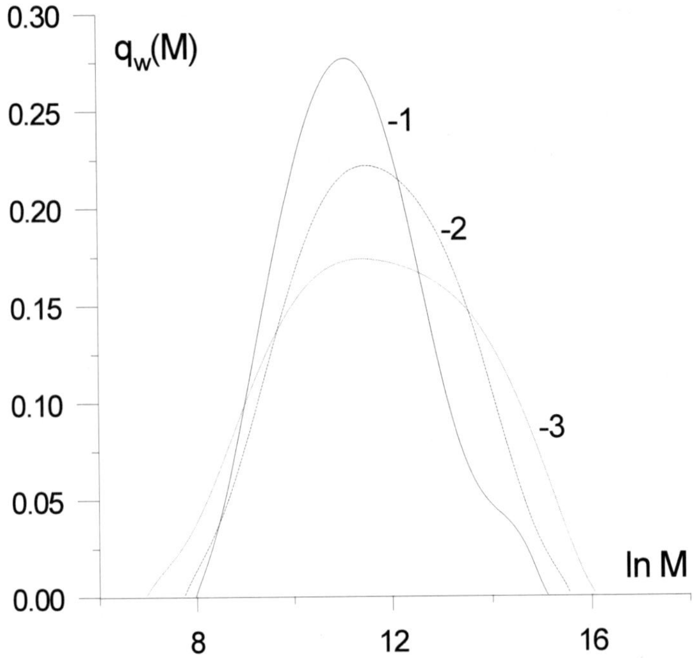

Fig. 3. Molecular weight distribution curves for polybutadiene, received on $NdCl_3{\cdot}3L\text{-}Al(C_6H_{13})_3$ catalytic system. Polymerization conditions are given in Fig. 1. Monomer conversion (%): 1 – 1,7; 2 – 8,7; 3 – 89,4. M_w/M_n:1 – 4,8; 2 – 7,3; 3 – 17,8.

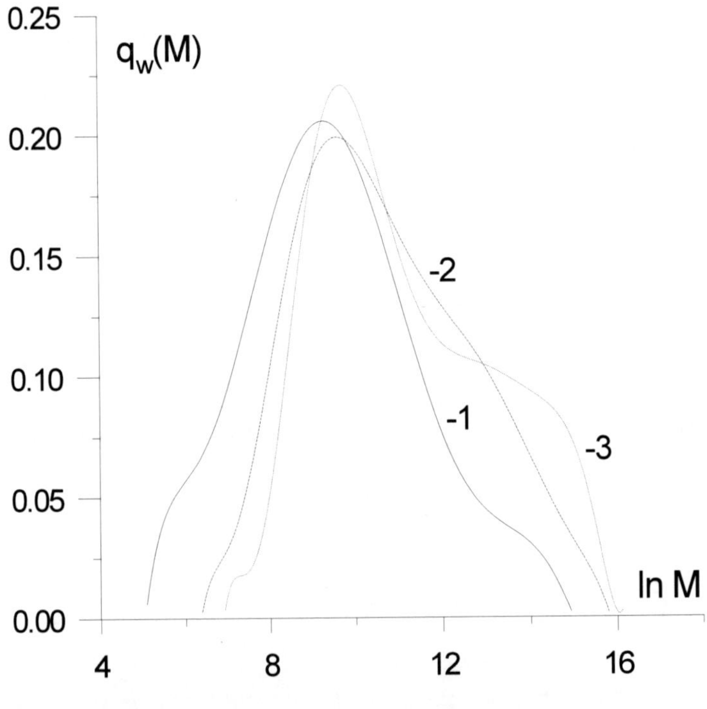

Fig. 4. Molecular weight distribution curves for polybutadiene, received on $NdCl_3{\cdot}3L\text{-}Al(C_8H_{17})_3$ catalytic system. Polymerization conditions are given in Fig. 1. Monomer conversion (%): 1 – 3,1; 2 – 21,6; 3 – 91,3. M_w/M_n:1 – 18,2; 2 – 18,6; 3 – 20,8.

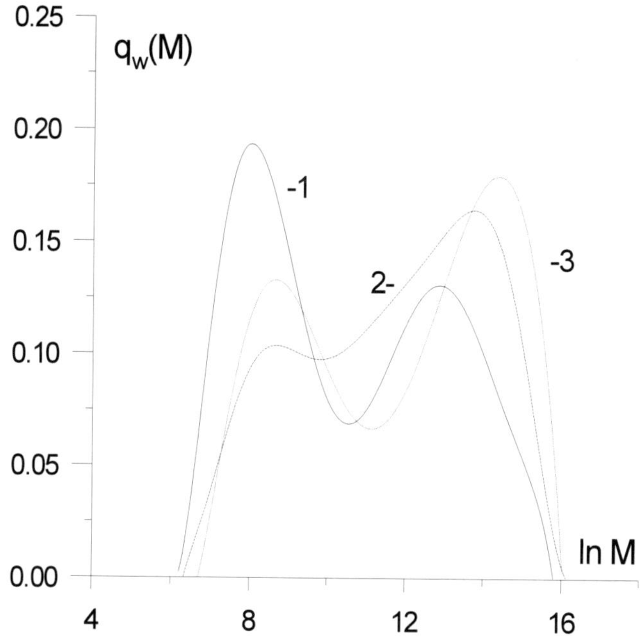

Fig. 5. Molecular weight distribution curves for polybutadiene, received on $NdCl_3 \cdot 3L\text{-}Al(C_2H_5)_3$ catalytic system. Polymerization conditions are given in Fig. 1. Monomer conversion (%): 1 – 3,2; 2 – 9,6; 3 – 88,0. M_w/M_n:1 – 28,3; 2 – 42,2; 3 – 55,2.

Distributions on kinetic activity of investigated catalytic systems $NdCl_3 \cdot 3L$-OAC were received with the help of Tikhonov regularization method on the base of MWD curves of polybutadiene obtained experimentally (Figures 6-9).

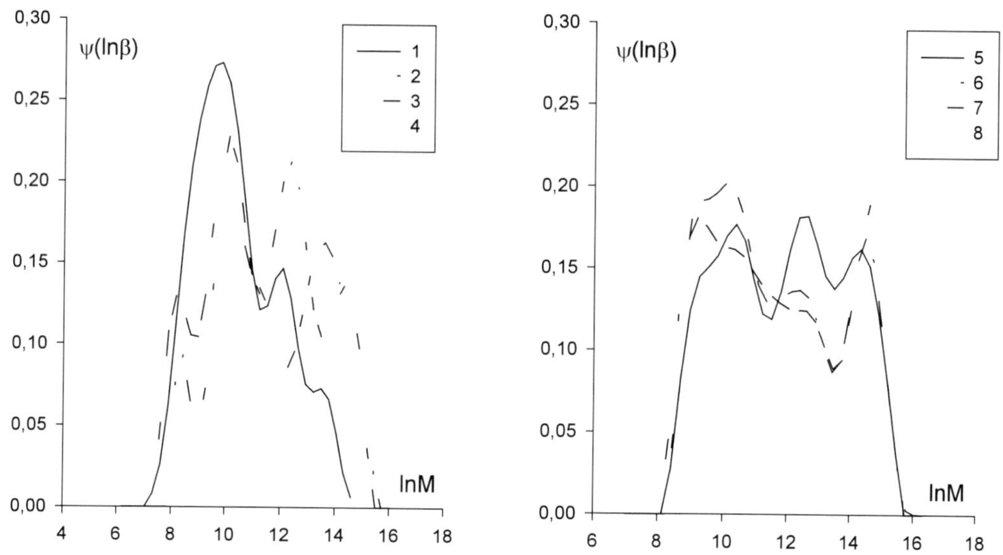

Fig. 6. Activity distributions of the $NdCl_3 \cdot 3L\text{-}Al(i\text{-}C_4H_9)_3$ catalytic system under butadiene polymerization; monomer conversions (%): 1 – 3, 61; 2 – 11,1; 3 – 17,4; 4 – 25,6; 5 – 55,6; 6 – 67,8; 7 – 87,9; 8 – 88,8.

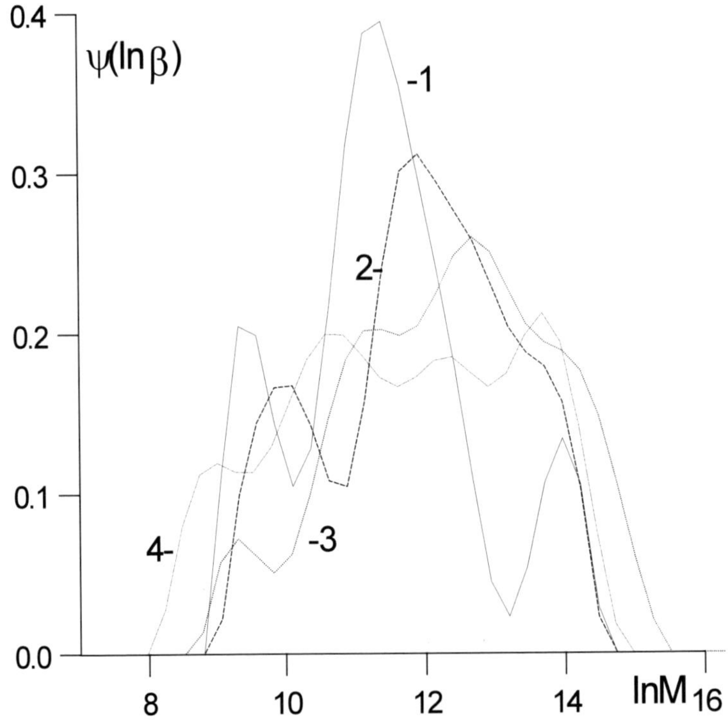

Fig. 7. Activity distributions of the $NdCl_3 \cdot 3L\text{-}Al(C_6H_{13})_3$ catalytic system under butadiene polymerization; monomer conversions (%): 1 – 1,7; 2 – 8,7; 3 – 26,2; 4 – 72,0.

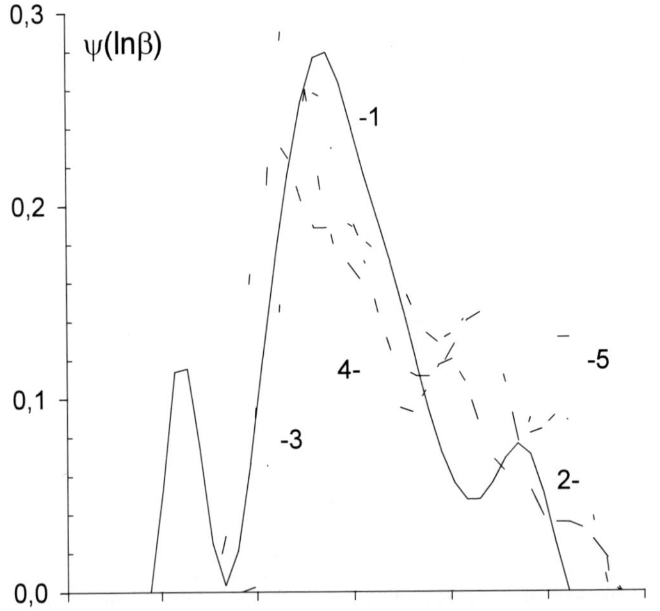

Fig. 8. Activity distributions of the $NdCl_3 \cdot 3L\text{-}Al(C_8H_{17})_3$ catalytic system under butadiene polymerization; monomer conversions (%): 1 – 3,1; 2 – 13,4; 3 – 21,6; 4 – 49,6; 5 – 57,9.

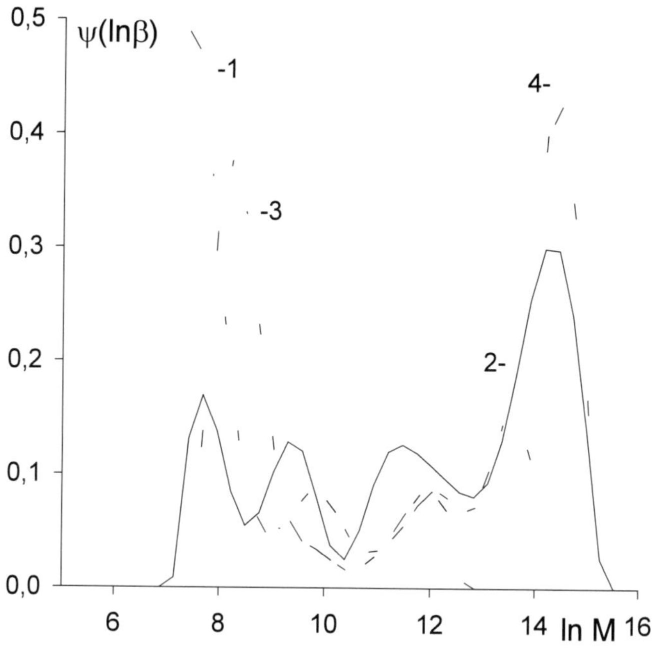

Fig. 9. Activity distributions of the $NdCl_3{\cdot}3L$-$Al(C_2H_5)_3$ catalytic system under butadiene polymerization; monomer conversions (%): 1 – 3,2; 2 – 9,6; 3 – 40,2; 4 – 88,0.

All curves of distribution on kinetic activity are polymodal independently from the nature of using OAC, and four greatly expressive maximums appear with the conversion degree increase. It testifies that polymerization process occurs on at least four types of active centers for all investigated systems. More over, multimodality of $\psi(ln\beta)$ curves is obtained even at very small monomer conversions. It indicates that different types of active sites are formed immediately, at the initial stage of the process, i.e. catalytic system is primordially kinetically heterogeneous.

Every point on distribution curves characterizes the active centers' part with the chain termination probability equal β_l and produces macromolecules with molecular mass M_l. The fact of the presence of $\psi(ln\beta)$ distribution itself, testifies about kinetic heterogeneity of catalytic system.

Distributions on kinetic activity of catalytic system $NdCl_3{\cdot}3L$-$Al(C_2H_5)_3$ (Figure 9) differ from the $\psi(ln\beta)$ curves, received in the presence of $Al(i\text{-}C_4H_9)_3$, $Al(C_6H_{13})_3$, and $Al(C_8H_{17})_3$. For the system with $Al(C_2H_5)_3$ the presence of two clear maximums is characteristic. They correspond to the active centers that produce the very low molecular and the very high molecular polymer fractions. These two active centers types produce the bulk of polymer and the polymerization product fractions produced on every of active centers greatly differ in molecular mass – low molecular fraction $10^3 \div 10^4$ and high molecular fraction $\sim 10^6$. The contribution of active centers that form macromolecules with intermediate values of molecular mass is not significant. Apparently, just in consequence of such unequivalent action of active centers of $NdCl_3{\cdot}3L$-$Al(C_2H_5)_3$ catalyst, the formed polybutadiene has bimodal MWD.

The received distributions may be divided into separate Gaussian functions, assuming that the deviations of active center from the maximum are statistical (Figure 10). In this case the area of every Gaussian curve will correspond to the numerical part of polymer, produced on the corresponding center. This allowed us to estimate a relative contribution of each active center type and to follow their changes.

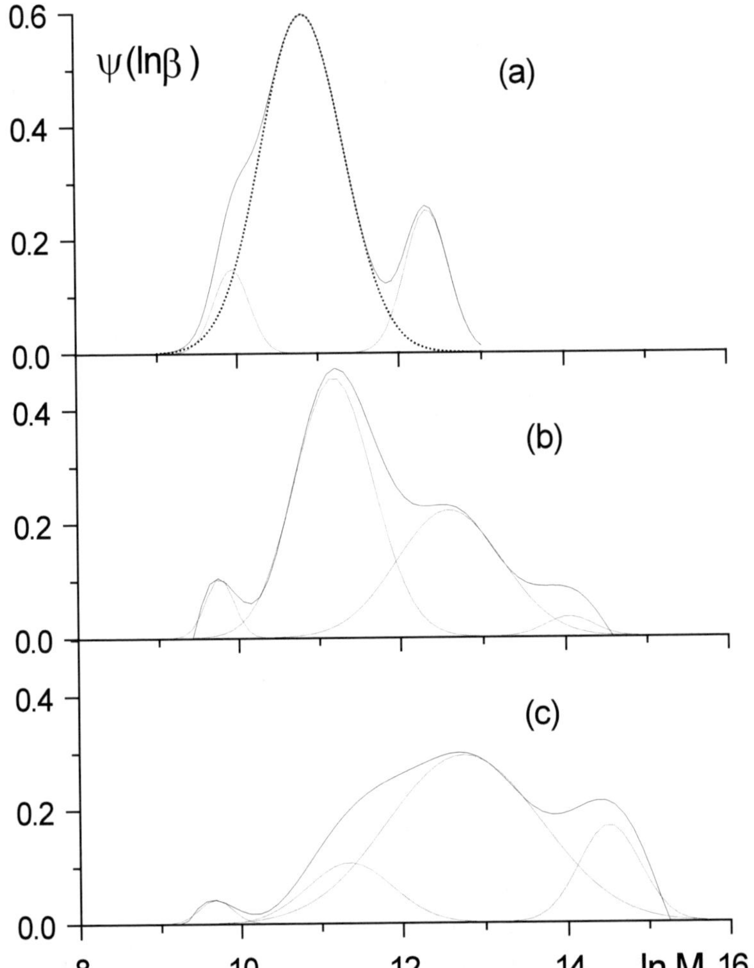

Fig. 10. Activity distributions of the $NdCl_3 \cdot 3L\text{-}Al(i\text{-}C_4H_9)_3$ catalytic system divided into separate peaks; monomer conversions (%): a – 2,6; b – 7,5; c – 53,4.

Figure 11 presents the change of maximums' positions, received by dividing the curve of kinetic activity distribution of catalytic system $NdCl_3 \cdot 3L\text{-}Al(i\text{-}C_4H_9)_3$ into separate peaks, in dependence on monomer conversion. One can say that maximums' positions corresponding to the various types of active centers remain virtually unchanged throughout the polymerization process. That is, each active center seems to synthesize macromolecules of a certain length.

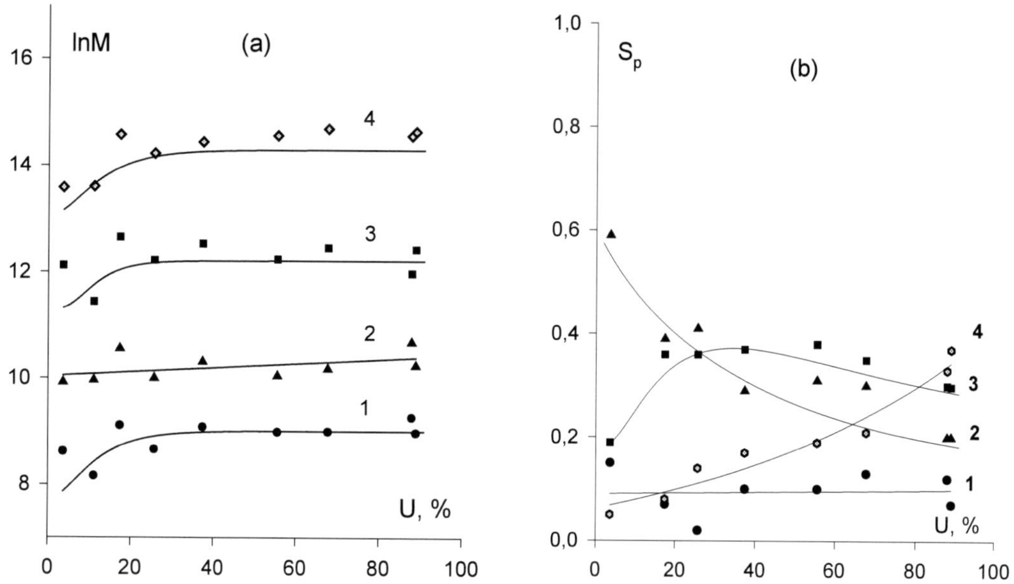

Fig. 11. Positions of maximums (a) and kinetic activity (b) of butadiene polymerization centers vs. monomer conversion. $NdCl_3{\cdot}3L$- $Al(i\text{-}C_4H_9)_3$ catalytic system.

Besides, active centers of catalytic systems with $Al(i\text{-}C_4H_9)_3$, $Al(C_6H_{13})_3$ and $Al(C_8H_{17})_3$ produce fractions with identical molecular mass. Thus, the active center, that produce most high molecular fraction forms polymer with molecular mass ~10^6, the first active center produces lowest molecular mass – $7 \div 8{\cdot}10^3$, the second and the third – $2{\cdot}10^4$ and $1{,}5{\cdot}10^5$, correspondingly. For the system with $Al(C_2H_5)_3$, position of the first (low molecular) peak slightly differs, the corresponding active center produces polymer with molecular mass ~$3{\cdot}10^3$, i.e. product with low molecular mass, comparing with catalysts on the base of $Al(i\text{-}C_4H_9)_3$, $Al(C_6H_{13})_3$ and $Al(C_8H_{17})_3$.

The activity of centers determined by the area of this maximum on the curve of kinetic activity distribution is changed throughout the polymerization process. In Figure 12 there are the changes of kinetic activity of every active center under the butadiene polymerization process on the catalytic system $NdCl_3{\cdot}3L$-$Al(C_2H_5)_3$. It is obvious that active centers generating the lowest molecular fractions and corresponding to the first maximum on the distribution curve, are characterized by the very low activity, which is slightly increased in the course of polymerization.

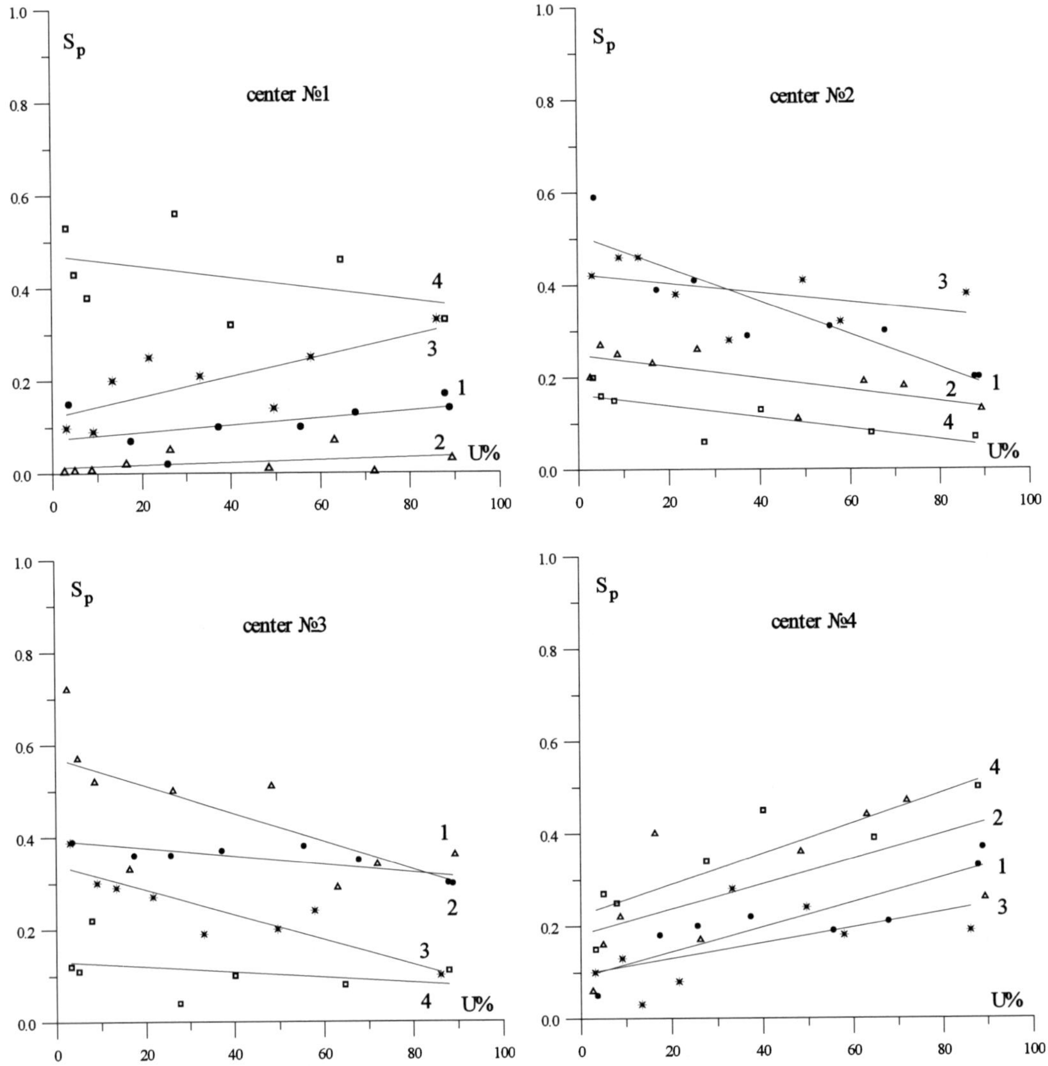

Fig. 12. The plots of kinetic activity of four types of butadiene polymerization active centers vs. monomer conversion. $NdCl_3 \cdot 3L$-OAC catalytic system. (1 – $Al(i\text{-}C_4H_9)_3$, 2 – $Al(C_6H_{13})_3$, 3 – $Al(C_8H_{17})_3$, 4 – $Al(C_2H_5)_3$). Polymerization conditions are given in Fig. 1.

Activity reduce is obtained for the second active center type, the growth with saturation – for the third, and constant activity increase throughout the polymerization is obtained for the fourth type. Some disconnection of data is explained by the static nature of polymerization process. It is really difficult to prepare two polymer products with equal molecular masses and consequently, with identical MWD in two different experiments. In its turn, even insignificant deviations in MWD form lead to the essential changes in distributions on kinetic activity. However, one can estimate the tendency of the activity changes of various centers. Curves of the kinetic activity change of $NdCl_3 \cdot 3L$-OAC catalytic systems are separately grouped for the every active center's type in Figure 13. It permits to establish the tendency of kinetic heterogeneity change of catalytic systems, containing different organoaluminum compound. It is obvious, that under the use of $Al(i\text{-}C_4H_9)_3$, $Al(C_6H_{13})_3$ and $Al(C_8H_{17})_3$ as organoaluminum compound kinetic activity of all four centers type is different in intensity, but is equally changed with monomer conversion. Active centers generating the lowest molecular fractions increase their activity in the course of polymerization. The activity of this center is decreased for the catalytic system with the use of $Al(C_2H_5)_3$. Activity of centers corresponding to the second and to the third maximums decreases for all used catalytic systems. There is the constant growth for the fourth type of active center, independently from the using in catalytic system organoaluminum compound. The low values of intensities of the second and the third peaks for $NdCl_3{\cdot}3L\text{-}Al(C_2H_5)_3$ catalytic system also must be noted.

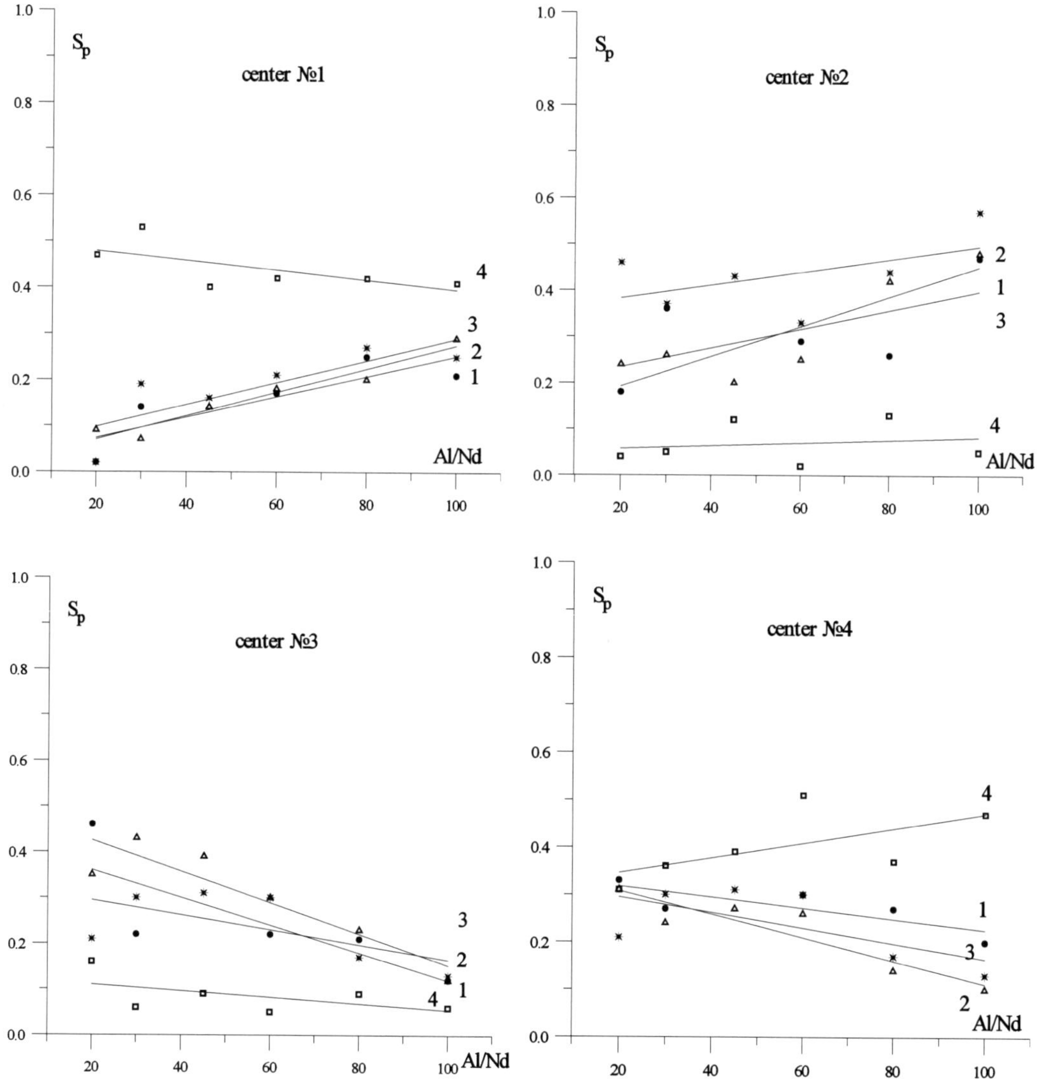

Fig. 13. Kinetic activity change of butadiene polymerization centers at various ratio Al/Nd. $NdCl_3{\cdot}3L$-OAC catalytic system (1 – $Al(i\text{-}C_4H_9)_3$, 2 – $Al(C_6H_{13})_3$, 3 – $Al(C_8H_{17})_3$, 4 – $Al(C_2H_5)_3$).

The kinetic activity change of four active centers types of investigated catalytic systems at various ratios Al/Nd is depicted in Figure 14. It is obvious, that the course of kinetic activity dependence on Al/Nd ratio practically coincides under the use of $Al(i\text{-}C_4H_9)_3$,

$Al(C_6H_{13})_3$ and $Al(C_8H_{17})_3$, as in the case of conversion curves. There are the increases of activity for the first and the second peaks and activity decrease is for the third and the fourth peaks with the growth of OAC content in catalytic system. In contrast to other investigated catalysts, for the catalytic system $NdCl_3{\cdot}3L\text{-}Al(C_2H_5)_3$ the first maximum has high activity values and slightly reduces with the increase of Al/Nd ratio. The change of the fourth peak intensity also differs from the rest catalytic systems. With $Al(C_2H_5)_3$ content growth its intensity increases.

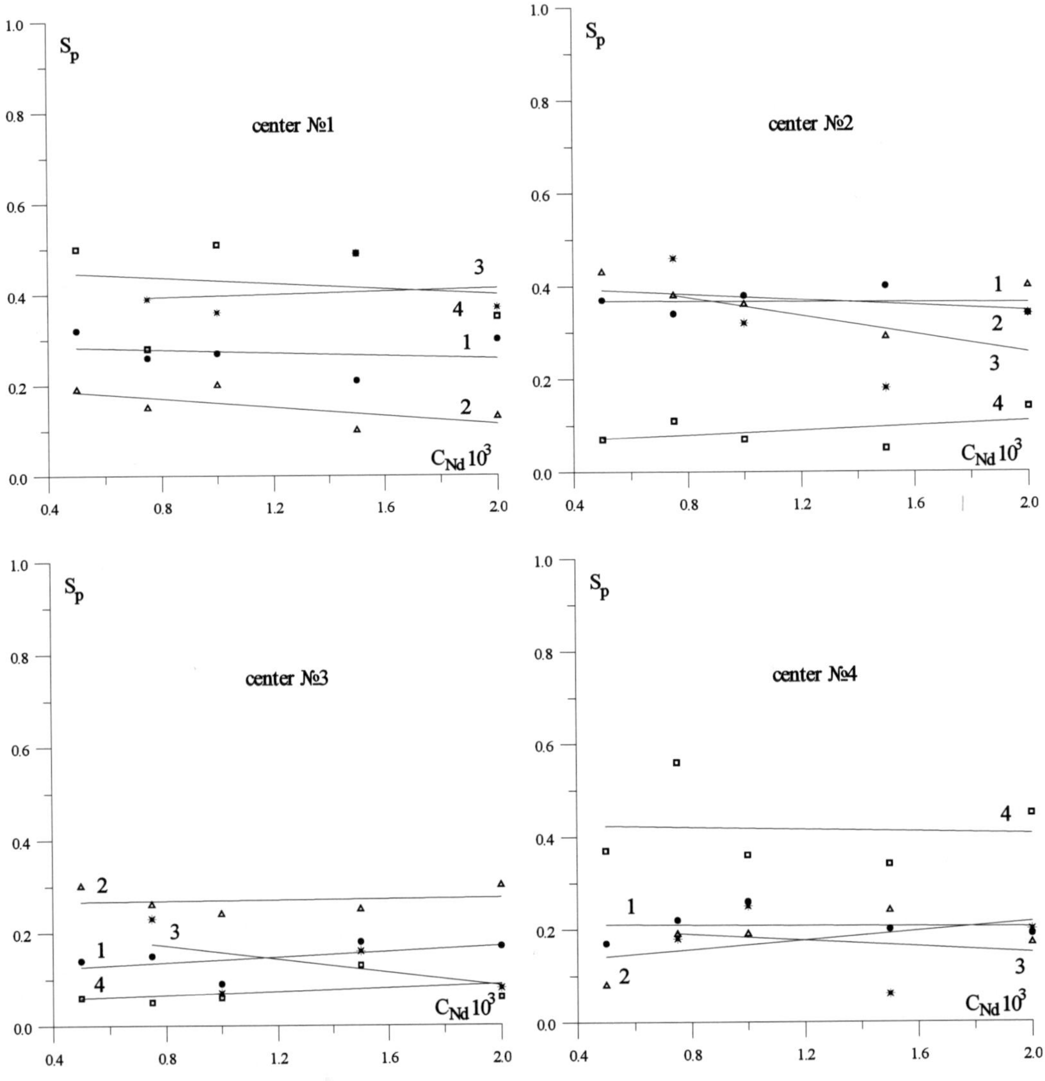

Fig. 14. Kinetic activity change of butadiene polymerization centers at various catalyst concentrations. $NdCl_3{\cdot}3L$-OAC catalytic system (1 – $Al(i\text{-}C_4H_9)_3$, 2 – $Al(C_6H_{13})_3$, 3 – $Al(C_8H_{17})_3$, 4 – $Al(C_2H_5)_3$).

As follows from Figure 15, the change of catalyst concentration in polymerization systems practically doesn't lead to the change of kinetic activity of polymerization centers.

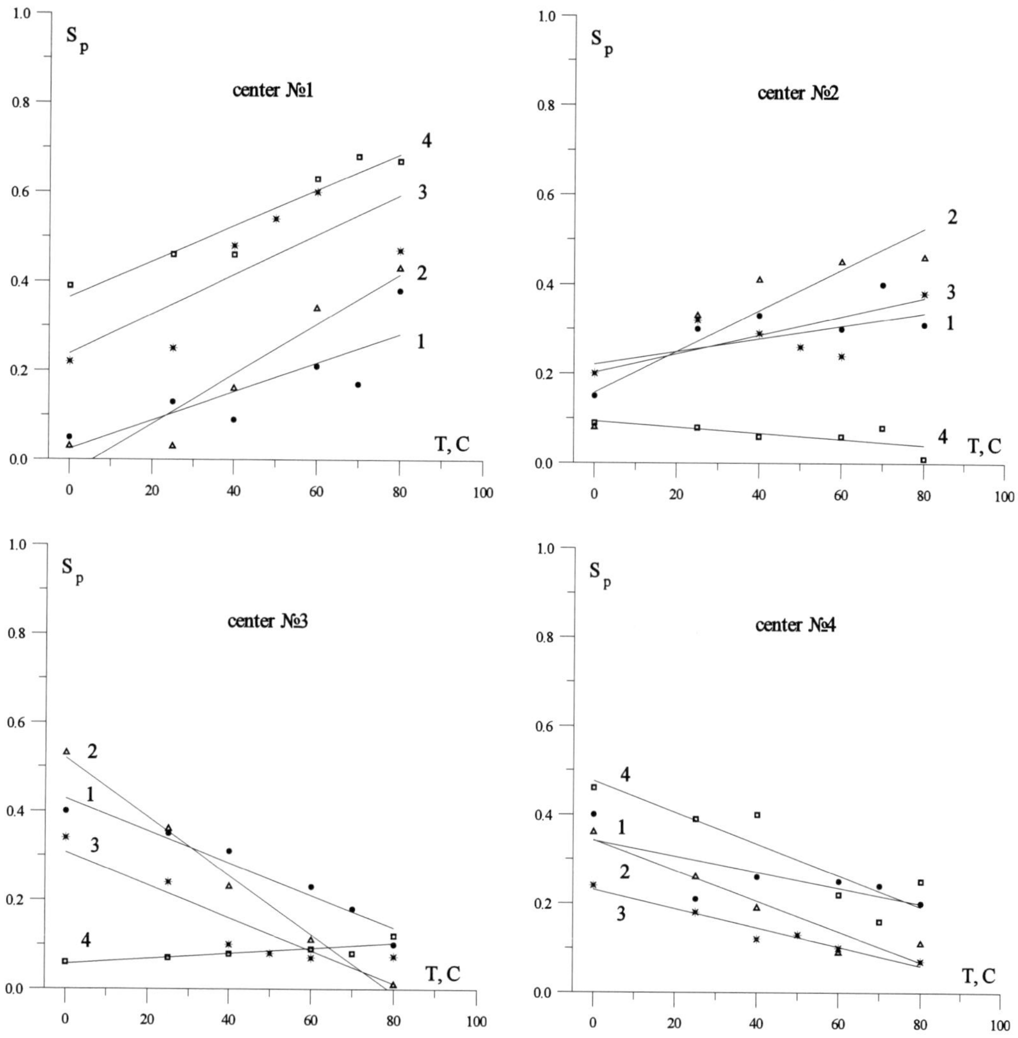

Fig. 15. Kinetic activity change of butadiene polymerization centers at various polymerization temperatures. $NdCl_3 \cdot 3L$-OAC catalytic system (1 – $Al(i\text{-}C_4H_9)_3$, 2 – $Al(C_6H_{13})_3$, 3 – $Al(C_8H_{17})_3$, 4 – $Al(C_2H_5)_3$).

The influence of butadiene polymerization temperature on the distributions of kinetic activity of investigated catalytic systems was also studied (Figure 16). It is obvious, that with the temperature increase the intensity of the third and the second peaks, producing low molecular polymer, grows. Opposite, centers responsible for generating of high molecular fractions (peaks 3 and 4) decrease their activity. Obviously, active centers forming high molecular fractions are less thermostable than active centers, on which the low molecular polymer is formed. As for conversion dependences the kinetic activity of centers corresponding peaks 2 and 3 is very low when using $Al(C_2H_5)_3$ as organoaluminum component.

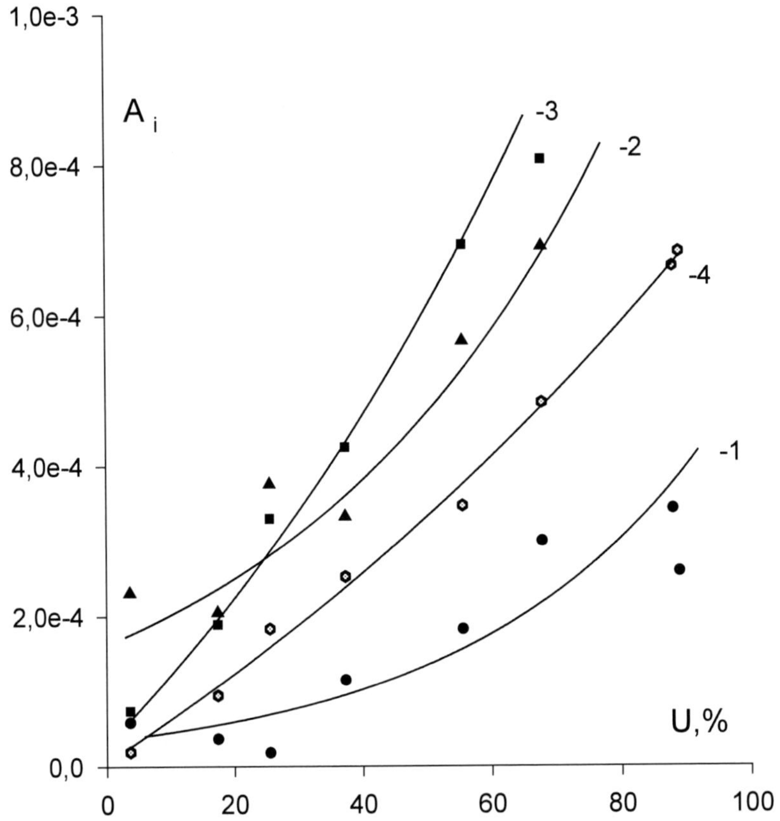

Fig. 16. The plot of A_i vs. monomer conversion under butadiene polymerization on $NdCl_3{\cdot}3L$- $Al(i\text{-}C_4H_9)_3$ catalytic system.

The magnitude S_i – area of peak – was received by dividing the activity distributions into separate Gaussian functions and, as it has been said earlier, characterizes the part of chains formed on the *i* active center. S_i is equal the ratio of the quantity of chains received on *i* active center to the overall chains number Q_0:

$$S_i = \frac{Q_{oi}}{Q_o} \tag{5}$$

Suppose

$$\mu = C_{m0} \frac{C_{ai}}{\sum C_{ai}} x \tag{6}$$

is the part of monomer expended on *i* active center at conversion moment *x*, where C_{ai} – *i* center concentration.

Then, the magnitude

$$\frac{k_{mi}C_m + k_{Ali}C_{Al}}{k_{pi}C_m}\mu \tag{7}$$

is the number of "dead" chains, formed on i center by transfer reactions to monomer and OAC at conversion moment x, where k_{mi} and k_{Ali} are the rate constants for the reactions of chain transfer to the monomer and OAC, accordingly, k_{pi} – is the rate constant of chain propagation on the corresponding type of active centers.

The total number of macromolecules produced on the i center is:

$$Q_{oi} = C_{ai} + \frac{k_{mi}C_m + k_{Ali}C_{Al}}{k_{pi}C_m} \cdot \frac{C_{m0}C_{ai}}{\sum C_{ai}} x \tag{8}$$

where the first item C_{ai} characterizes the number of "living" chains.

The total number of chains is determined by the ratio of the quantity of reacted monomer and the number-average degree of polymerization P_n:

$$Q_o = \frac{C_{m0}}{P_n} x \tag{9}$$

With the regard for (8) and (9) after noncomplicated transformations expression (5) may be represented as:

$$\frac{S_i C_{m0} x}{P_n} = C_{ai}\left[1 + \frac{C_{m0}k_{mi}}{k_{pi}\sum C_{ai}} x + \frac{C_{Al}k_{Ali}C_{m0}}{C_m k_{pi}\sum C_{ai}} x\right] \tag{10}$$

Then let's assume:

$$\frac{S_i C_{m0} x}{P_n} = A_i$$

The position of maximum of the curve of distribution on kinetic activity characterizes the polymerization degree P_{ni} of polymer fractions, formed on the corresponding active centers. The polymerization degree P_{ni}, in its turn, is determined as a ratio of the monomer quantity μ, reacted on i center and number of chains Q_{0i}, formed on it.

Using expression (6), one receives:

$$P_{ni} = \frac{C_{m0} C_{ai}}{Q_{0i} \sum C_{ai}} x \tag{11}$$

Taking into account that conversion is determined as

$$x = 1 - e^{-\tau \Sigma k_{pi} C_{ai}}$$

for the initial polymerization moment, at $\tau \to 0$, one can assume:

$$x \approx \tau \Sigma k_{pi} C_{ai} = \tau \langle k_p \rangle \Sigma C_{ai} \tag{12}$$

where $\langle k_p \rangle$ is a generalized constant of propagation rate. With regard for (12) the equation (11) may be represented as:

$$P_{ni} = C_{m0} \langle k_p \rangle \tau \tag{13}$$

As it follows from equation (13) the angle of dependences of P_{ni} on τ at $\tau \to 0$ is constant magnitude for all active centers and characterizes this catalytic system. Thus, the maximums' positions shift in Figure 10 at the initial moment of polymerization is a natural phenomenon, and this effect must be observed in all cases under more detailed study of polymerization process at low conversions.

One can receive the following expression for analysis P_{ni} by substituting (8) in (11):

$$B_i = \frac{C_{mo} C_{ai} x}{P_{ni} \sum C_{ai}} = C_{ai} + \frac{k_{mi} C_m + k_{Ali} C_{Al}}{k_{pi} C_m} \cdot \frac{C_{m0} C_{ai}}{\sum C_{ai}} \tag{14}$$

Expressions (10) and (14) permit to analyze kinetic activity change of catalytic systems more clearly. The dependences of magnitude A_i on monomer conversion x are depicted in Figure 15. Extrapolation of this curves to zero conversion, according to equations (10) and (14) makes possible to determine C_{ai} – the concentration of every active centers separately, and also the total concentration of polymerization centers ΣC_{ai}.

The values of concentrations of different active centers types under butadiene polymerization on investigated catalytic systems are represented in the table. ΣC_{ai} – the total concentration of all the active centers' types; values $\langle C_a \rangle$ are also the total active centers concentration, but they were received by analysis of polymerization process kinetics by the equation:

$$W_p = k_p C_m \langle C_a \rangle$$

It is obvious, that there is good coincidence of results of the both calculation methods, that confirms adequateness of the used model.

Table. The neodymium catalytic systems active centers concentrations under butadiene polymerization

Catalytic system	$C_{ai}\,10^5$(mol/l)				$\Sigma C_{ai}\,10^5$	$<C_a>10^5$
	C_{a1}	C_{a2}	C_{a3}	C_{a4}	mol/l	mol/l
$NdCl_3{\cdot}3L\text{-}Al(i\text{-}C_4H_9)_3$	3,9	21,7	7,5	5,7	38,8	40,1
$NdCl_3{\cdot}3L\text{-}Al(C_6H_{13})_3$	0,05	0,9	1,5	0,8	3,3	3,0
$NdCl_3{\cdot}3L\text{-}Al(C_8H_{17})_3$	2,39	6,2	6,9	28,7	65,6	65,0
$NdCl_3{\cdot}3L\text{-}Al(C_2H_5)_3$	6,8	3,9	3,0	6,8	20,5	25,0

CONCLUSION

Thus, heterogeneity of investigated catalytic systems has been displayed in the existence of four types of active centers. They are formed at the beginning of polymerization process and produce macromolecules with the lengths, which are definite for each type of active centers. Kinetic activity of polymerization centers depends on OAC concentration and nature, and polymerization temperature. For the catalytic systems with $Al(i\text{-}C_4H_9)_3$, $Al(C_6H_{13})_3$ and $Al(C_8H_{17})_3$ kinetic activity of polymerization centers equally changes in dependence on polymerization conditions and the change of monomer conversion. Maybe, this is the consequence of identical structure of different active centers types of these catalytic systems.

The difference of distributions on kinetic activity and behavior of active centers of catalytic system with $Al(C_2H_5)_3$ with similarity in behavior of systems with $Al(i\text{-}C_4H_9)_3$, $Al(C_6H_{13})_3$ and $Al(C_8H_{17})_3$ can be explained by the difference in association degree of free OAC molecules in polymerization system. It is known that organoaluminum compounds with direct carbon chain are dimeric. Inclination to association is greatly increased with the decrease of carbon radical length. Thus, dimmer molecules of $Al(C_2H_5)_3$ are stable even in vaporous state and dissociate only at temperature more than 100°C. In solutions $Al(C_2H_5)_3$ takes the monomer shape only under dilution to more than 0,1% concentration. Probably, the bigger inclination to association of $Al(C_2H_5)_3$ molecules, in comparison with $Al(C_6H_{13})_3$, $Al(C_8H_{17})_3$ and $Al(i\text{-}C_4H_9)_3$ ($Al(i\text{-}C_4H_9)_3$ is incapable of association, because it has the ramified structure of carbon radical) leads to the obtained differences in distributions of polymerization centers.

The method of analysis of curves' maximums of distribution on kinetic activity and the change of kinetic activity of every active center's type was proposed. It permits to calculate the values of concentration for every type of active centers separately.

REFERENCES

[1] V.G. Kozlov, N.G. Marina, I.G. Savel`eva, Yu.B. Monakov, G.A. Tolstikov, *Inor. Chim. Acta.*, **154**, 239 (1988).

[2] V.G. Kozlov, N.N. Sigaeva, K.V. Nefedjev, I.G. Savel'eva, N.G. Marina, Yu.B. Monakov, *J. Polym. Sci., Polym. Chem.*, **32**, 1237 (1994).

[3] H.L. Hsien, G.H.C. Yeh, *Rubber Chem. and Tachnol.*, **58**, 117 (1985).

[4] Yu.B. Monakov, N.G. Marina, R.M. Khairullina, O.I. Kozlova, G.A. Tolstikov, *Inor. Chim. Acta.*, **142**, 161 (1988).

[5] N.N. Sigaeva, T.S. Usmanov, E.A. Shirokova, V.P. Budtov, S.I. Spivak, Yu.B. Monakov, *Dokl. Phys. Chem.*, **365**, 83 (1999).

[6] N.N. Sigaeva, T.S. Usmanov, V.P. Budtov, S.I. Spivak, Yu.B. Monakov, *Polym. Sci. B.*, **42**, 753 (2000).

[7] Yu.I. Murinov, Yu.B. Monakov, Z.G. Shamaeva, N.G. Marina, V.S. Kolosnitsyn, Yu.E. Nikitin, S.R. Rafikov, *Izv. Akad. Nauk SSSR Seriya himich.*, **12**, 2790 (1977).

[8] H. Benoit, L. Crubisic, P. A. Rempp, *J. Polym. Sci. B.*, **5**, 753 (1967).

[9] A.A. Berg, V.G. Kozlov, V.P. Budtov, Yu.B. Monakov, S.R. Rafikov, *Vysokomol. Soed.*, **22**, 543 (1980).

[10] T. Provder, E.M. Rozen, *Separ. Sci.*, **5**, 437 (1970).

[11] A.N. Tikhonov, A.V. Goncharsky, V.V. Stepanov, and A.G. Yagola, *Numerical Methods for the Solution of Ill-Posed Problems*, Kluwer, Dordrecht, 1995, ISBN 0-7923-3593-X.

[12] I.V. Kochikov, G.M. Kuramshina, A.G. Yagola, *Surv. Math. Ind.*, **8**, 63 (1998).

[13] A.G. Yagola, *USSR Comput. Math. Math. Phys.*, **20**, 586 (1980).

[14] A.N. Tikhonov, A.S. Leonov, and A.G. Yagola, *Nonlinear Ill-posed Problems*, Chapman & Hall, 1998, ISBN 0-412-78660-5

[15] W.R. Schmeal, J.R. Street, *J. Polym. Sci., Polym. Phys. Ed.*, **10**, 2173 (1972).

Chapter 10

SPECIFIC FEATURES OF THERMOOXIDATION OF THERMORESISTANT HETEROCHAIN POLYMERS

E. V. Kalugina[a], T. N. Novotortseva[a], K. Z. Gumargalieva[b] and G. E. Zaikov[c]

[a]Research and Production Company "POLYPLASTIC", 14a, General Dorokhov str., Moscow, 119530, Russia;

[b]Inst. of Chemical Physics, 4b, Kosygin str., Moscow, 117334, Russia;

[c]Inst. of Biochemical Physics, 4b, Kosygin str., Moscow, 117334, Russia.

The main progress in the science of aging and stabilization of polymers generally relate to large-tonnage carbochain polymers: polyolefins, poly(vinyl chloride), rubbers, acrylic polymers. This is defined not only by demands to the industry, but also by the presence of the fundamental investigation of cracking and oxidation of individual aliphatic compounds.

No such investigations were performed for aromatic heterochain polymers (PHA). Thus, there are many unclear and doubtful aspects in understanding the degradation mechanism.

Works, performed in RIPM and SIE "Polyplastic" during last 15 years concerning PHA aging and stabilization, were analyzed. Specific features of high-temperature (300 – 400°) oxidation in melt and solid-phase (150 – 200°) oxidation of PHA.

General regularities in thermooxidation of aromatic and aliphatic-aromatic polyimides, polyesterimides, poly(phenylquinoxalines), polysulfones, etc. are brought out.

It is shown that PHA degrades according to the radial-chain mechanism, but the process differs from that of aliphatic polymers by the presence of initiation acts. Thermally realized electron-excited states of polymers and molecular complexes with oxygen are shown up as the most probable structures.

Possible directions of PHA stabilization with the help of fluorine-containing additives and/or compounds of variably valence metals are discussed. At high temperature, degradation changes in polyimides, based on dianhydrides of diphenyltetracarboxylic and benzophenontetracarboxylic acids, m- and p-phenylene diamines, aniline fluorene, poly(alkane imides), etc., reduce by 1.5 – 2 times. For example, a glass-plastic composite based on stabilized polyimide linkage preserves strength no lesser than 150 MPa at 350° during 1000 hours, therewith an unstabilized material degrades completely during this time.

Beside deceleration of thermooxidative aging, phosphorus-containing additives increase radiation resistance of, for example, polyimides and poly(alkyl terephthalates): the extraction rate of aromatic rings radiation destruction CH_4-product two-fold reduces.

1. THERMOOXIDATIVE DEGRADATION OF POLYPYROMELLITIMIDE, POLY(PHENYL-QUINOXALINE) AND COPOLY(IMIDOPHENYLQUINOXALINES) [1 – 11]

Judging from the temperature index value, determined by the methodology of the research organization National Bureau of Standards, USA, temperature limits of aromatic polyimide (PI) of Capton trademark (Russian analogue is PM film), within which the material preserves 30% of exploitation properties during 30 thousand hours, is 200 – 220°. PM absorbs 0.5 mole of O_2/base-mole at 200° during 2000 hours of oxidation. Therewith, 0.2 mol/base-mole CO_2 is released. At 250°, kinetics of O_2 absorption displays all its specific features. The film absorbs 3.5 mole of O_2/base-mole during 5 thousand hours. Three stages are clearly displayed on kinetic curves:

- The first σ-idal stage lasts during about 1000 hours with absorption of 1.5 mol/base-mole O_2 and is described well by the first order equation;
- The second stage with exposure during up to 4000 hours displays almost constant rate of O_2 absorption (zero kinetic order);
- The third stage displays acceleration of the reaction, and kinetics is described by the equation $\chi = A\times\exp(\gamma^{-\tau})$, where $\gamma^{-\tau}$ is the autoacceleration factor, s^{-1}.

Therewith, the main gas products are H_2, CO, CO_2, and H_2O, which type of release kinetics is similar to O_2 absorption. Balance reduction by O_2, absorbed by PM and released with oxidation products, shows that on initial stages the polymer links 50% of O2. According to IR-spectroscopy data, new oxygen-containing structures of the quinoid type (absorption at 1640 cm^{-1}) occur in the oxidized film. Higher expanse of aromatic rings compared with heterocycle (comparison of absorption bands of 1500 cm^{-1} for $C=C_{ar}$, 1090, 1120 and 820 cm^{-1} for $C\text{-}H_{ar}$, 1244 cm^{-1} for $C_{ar}\text{-}O\text{-}C_{ar}$ with 1380 and 720 cm^{-1} for imide cycles) is also displayed. Moreover, a solid individual product, identified as pyromellite diimide (PDI) by NMR and SM, was also observed at PM thermooxidation. As the result of degradation at 300°, outputs of PDI and H_2 are 0.015 and $1.1\cdot10^{-3}$ mol/base-mole, respectively. Formation of H_2 explains PM cross-linking, in which gel-fraction increases at solid-phase aging: ~3 cross-links per macromolecule are formed at 300° during 20 hours. On the example of model compound, N-phenylphthalimide (PPI), oxidized under the same conditions, typicalness of the composition of PM oxidation products was proved, namely, the presence of PDI and H_2. Therewith, no PDI and H_2 are formed during thermolysis of PPI and PM.

Thermooxidation kinetics and composition of degradation products of poly(phenylquinoxaline) (PPQ) at 200 – 300° are analogous to PM. The following compounds were identified: H_2, CO, CO_2, H_2O, and 2,2'-(1,4-phenylene)-bis-(3-phenylpyrazine). Depending on exposure conditions, PPQ heating in the air causes partial or complete loss of polymer solubility. Heating at 300° during 20 hours formed about 4 cross-links per macromolecule. Gel-fraction accumulation correlates with H_2 release. New oxygen-containing structures of the quinoid type (IR-absorption at 1640 – 1670 cm^{-1}) are formed in PPQ. The authors have confirmed PPQ affinity to cross-linking in experiments on model compounds. For example, at thermooxidation under the conditions mentioned, beside carbon oxides and water, a dimer with molecular weight of 652 ($2M_{DPQ}$-2) and H_2, not formed under the same conditions in vacuum, were identified.

According to dynamic thermogravimetric analysis (TGA) data, thermostability of aromatic structure copolymers (PPQ-1) in tests in vacuum does not practically depend on concentration of ether-aromatic units. For PPQ-1, temperature corresponded to 10% of mass losses (T_{10}) falls within the range of 570 – 580°. Thermal degradation of fatty aromatic copolymers (PPQ-2) starts at lower temperature and falls within the range of 520 – 450° decreasing with CH_2-groups content growth. Substitution of ether-aromatic fragments by aliphatic ones not only decreases PPQ thermal stability, but also changes the decomposition type. Degradation transforms from practically one-stage process into multistage one: intensive decomposition at 420 – 500° is changed by abrupt deceleration, which is probably associated with spending less thermostable aliphatic parts and formation of more thermostable structure. Above 650° the process is accelerated again. In the presence of O_2, degradation reactions are sharply intensified – SPPQ-1 begins degrading at 490 – 500°, SPPQ-2 – at 500 – 400°, $T_{decomp.start}$ decreases with growth of aliphatic parts amount. Increase of methylene groups amount in SPPQ-2 leads to intensifying absorption of O_2 and therewith formed CO and CO_2. For SPPQ-2 with CH_2 contents of 5.9 and 0.8, effective activation energies, calculated from kinetic absorption curves for O_2, are 72 and 116 kJ/mol, respectively. Comparison of SPPQ and homopolymer – poly-2,2'-(1,4-phenylene)-6,6'-oxy-bis-(3-phenylquinoxaline), shows that copolymers containing ether-aromatic and aliphatic groupings in 5.2 and 0.8 wt.% concentrations, respectively, differ insignificantly from the homopolymer. For SPPQ, PPQ and PI, the general runs of kinetic curves of O_2 absorption and carbon oxides and water extraction are identical. Already on the first stage of oxidation SPPQ-2 links in its structure more than 65% of O_2 from the absorbed amount, i.e. much more than PI and PPQ. According to DTA data, reactions of the aliphatic part (300 – 420°) and aromatic rings (above 420°) oxidation were successfully extracted during dynamic thermooxidation of SPPQ-2. Expense of CH_2 for oxidation is confirmed by IR-spectroscopy data – by narrowing the absorption band at 2900 cm^{-1} (ν_{CH_2}) keeping main absorption bands of heterocycles (1700, 1670, 1740, 1330, 1220 cm^{-1}). The presence of aliphatic fragments in the structure as if activates thermooxidation of heterocycles reducing the decomposition onset temperature by 20 – 60°. Oxidation of SPPQ, as well as homopolymers, is accompanied by formation of intermolecular bonds, the presence of methylene chains accelerating cross-linking. Gel-fraction concentration in SPPQ-2 becomes twice higher, than in PPQ at 300° during 20 hours of thermooxidation.

Thus, let us denote general regularities of PI, PPQ and SPPQ thermooxidation. The first important likeness of the degradation behavior is O_2 absorption at rather low temperature (200°). At temperatures, when it becomes possible to search O_2 absorption kinetics in a real time (250° or higher), polymers display equal oxidation type independently of temperature – the three-stage kinetics, sequentially described by laws of the first and the zero orders with future autoacceleration.

Only about 10% of absorbed O_2 are removed from polymers with carbon oxides and water on the first stage of oxidation. This effect failed to explain by difference in chains (structure defectiveness) of macromolecules, because PPQ possessing practically defectless structure absorbs almost twice more O_2, than PM film. E.g. there is practically no alternative to the statement that the first stage of PM, PPQ and SPPQ thermooxidation is associated with the features of their chemical structure. The second important likeness in the degradation behavior of the polymers mentioned (with account for structure of the elementary unit) is formation of a similar type of degradation products, namely, PMD and PPP. The ratio of PMD and CO_2 outputs (in moles) during PM thermooxidation at 300° is close to 1/24, i.e. to separate PMD, aromatic structures of amine components must be oxidized to CO_2. A random O_2 attack along the macrochain would cause formation of a selection of oligomeric products with end PDI-groups, but this does not happen. Intermolecular cross-linking, relatively less frequent (M_c = 1000) but inducing practically complete loss of polymer solubility, is general for PI and PPQ. Oxygen participates indirectly in cross-linking and is absent in the cross-linking unit (bridge). However, cross-linking in the absence O_2 does not practically proceed at these temperatures. Interaction of O_2 with an aromatic structure probably activates C-H-bonds break in the aromatic ring, which thereafter enter intermolecular interaction. Basing on the results of isotopic O_2 composition analysis under PM and PPQ thermooxidation under the condition of 30 – 50% O_2 conversion (high-temperature oxidation at 350 – 400°) it was concluded that a direct interaction between PM and PPQ with O_2 exists, and oxygen in the O-O form is present, i.e. so-called endoperoxides with aromatic rings are formed. Oxygen is easily attached to the triplet aromatic system (long wave limit of the fluorescence for PM is 520 – 530 nm, which corresponds to energy of 217 – 226 kJ/mol). Bond energies in model compounds were calculated with the help of quantum chemistry method (the Pariser-Parra-Pople method) and indicated the possibility of the mentioned reaction proceeding for both PI and PPQ. As endoperoxide is formed, π-electron conjugation system on the aromatic ring changes. Decomposition of endoperoxides leads to occurrence of oxygen-containing structures in the phenylene fragment. Deeper oxidation causes disclosure of the benzene ring and CO_2 and CO release. Electron excitation acts and formation of endoperoxides are just initial initiation stages. Paramagnetism increases in polymer with degradation as hexadiene oxidized structures, radicals and biradicals are accumulated, and the conjugation system increases. This is the third oxidation stage. This stage of the process is of theoretical interest only, because deterioration of exploitation characteristics of the materials ends already at the second stage.

The classical inhibition of oxidative processes is based on termination of oxidation kinetic chains and deactivation of branching intermediate products. In the case of PHA, the chain nature of the process does not seem obvious. Chain termination by additives suggests inertness of the remained inhibitor. These radicals are active at temperature above 200°. That is why, classical antioxidant are ineffective at high temperature. E.g. PHA should be thermostabilized in direction of reducing structure transition into the electron-excited state by quenching these states, interaction or complex formation with amine residues in macromolecules to reduce their reactivity. The highest effect in increasing thermostability was displayed by additives of phosphorus compounds (PCA). For example, at the optimal stabilizer content PDI output from PM reduces by 2.5 – 3 times, and output of carbon oxides by 1.5 – 2 times. Fiberglass plastic based on stabilized polyimide linkage preserves strength no less than 150 MPa during 1000 hours at 350°, and unstabilized material degrades completely during this time.

2. THERMOOXIDATIVE POLY(ALKANE IMIDE) DEGRADATION [12 – 29]

Poly(alkane imide) (PAI)

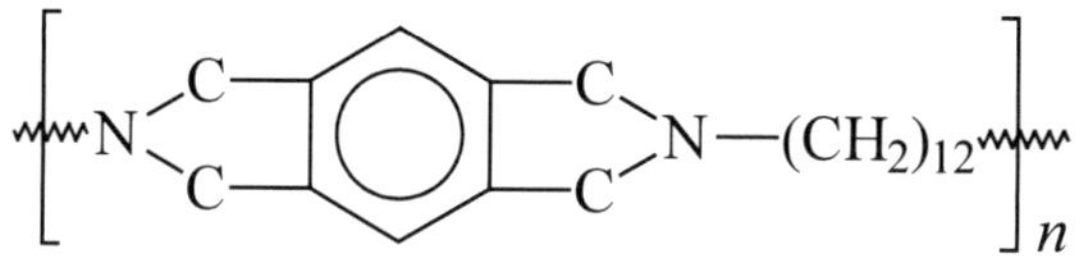

absorbs O_2 in the melt with release of CO_2, CH_2O, CO and H_2O. Kinetics of O_2 absorption at 300 – 350° is the two-stage process: the first is the fast stage with C-shaped kinetic curve, high initial rates and formally first kinetic order, and the second is the slow stage with a constant rate. Depending on temperature, the first stage lasts 2 – 5 hours with absorbing O_2 of about 8 mol/base-mol; the second stage lasts tens of hours with additional absorption of up to 2 moles of O_2. E_a of PAI oxidation in melt, determined by initial rates of O_2 absorption, equals 57 kJ/mol, whereas E_a of CO_2 and CO release (identical two-stage kinetics) equal 135 and 163 kJ/mol, respectively. CO_2 and CH_2O are the main by output light products of PAI high-temperature oxidation, e.g. their output in PAI oxidation at 350° during 1 hour equals 0.5 and 0.53 mol/base-mol, respectively. The output of CO and H_2O is much smaller. The balance by oxygen indicates that the main part of O_2 is linked by the polymer and is contained in heavy products.

Spectral data (NMR, Fourier-IR spectroscopy, mass-spectrometry) point to the destruction of methylene chains, expense of CH_2-groups and accumulation of aliphatic C=O-groups, destruction of imide cycles and occurrence of heavy products representing macrochain fractures at degradation of a fatty chain or heterocycle. For example, PDI, dodecamethylene diphthalimide, products of the R-PI-C(O)R' type, where R and R' are linear aliphatic radicals or H, and PI is the pyromellitimide residue. The part of products with destroyed imide cycles (phthalimides, nitrile phthalimides) increases at temperature above 350°. However, judging by output ratios of products with whole and partly degraded pyromellitimide cycles, high-temperature PAI oxidation is generally directed on fatty chains and oxidative degradation of cycloarylene structure is insignificant.

Judging by the shape of O_2 absorption and CO_2 and CO release kinetic curves, solid-phase PAI oxidation at 200 – 250° is similar to the initial stage of high-temperature oxidation, but it lasts hundreds of hours and is characterized by noticeably lower values of E_a = 21 kJ/mol. Although the balance by oxygen, absorbed and released with volatile products, points to its accumulation in the polymer, and IR-spectroscopy data confirm expense of CH_2-groups, accumulation of aliphatic C=O-groups and future low expense of imide cycles. Heavy oligomeric products are not extracted during solid-phase oxidation, excluding pyromellite diimide (PDI), which itself does not contain atmospheric oxygen and is just a fragment of the structure. Formally, aging kinetics is presented as complete oxidation of fatty chains to carbon oxides and extraction of pyromellitimide fragment, resistant to oxidation, as the product. Oxidation at 200° causes formation of hydroperoxides in PAI, accumulation kinetics of which is characterized by a maximum (~0.015 mmol/base-mol) at ~400 hour exposure. ROOH decay in vacuum at 200° proceeds due to the first kinetic order law with the rate constant of 0.017 min^{-1}. Judging by the value α (10^{-2}) characterizing ROOH output per absorbed oxygen, solid-phase oxidation in PAI develops mostly in the hydroperoxide direction. PAI thermooxidation is accompanied by H_2 release, which output after 800 hours of aging at 200° is 0.63 mmol/base-mol. Kinetics of H_2 release correlates with accumulation of gel-fraction at PAI aging.

At the first glance, phenomenology of high-temperature and solid-phase PAI oxidation is determined by fatty chain oxidation and must be described by the known radical-chain oxidation scheme of hydrocarbonic polymers with its modification accounting for a possibility of process development beside hydroperoxides, for example, via decay of peroxyradicals. The known attack reactions of O_2 on α-CH_2-bonds, weakened due to neighboring the π-conjugated system of the pyromellitimide fragment. The following experimental data exist, which do not allow exclusion of a cycloaromatic fragment from consideration of the PAI thermooxidation mechanism or reducing its role to activation of the neighboring methylene chain only, and at high temperature – participation in the conjugated oxidation:

- Methylene PAI chain is much more stable than aliphatic chains in PE and aliphatic ones in PA;
- PDI is the only heavy product of the solid-phase PAI oxidation;
- Correlation between PAI cross-linking and H_2 release at oxidation;
- PAI thermooxidation inhibition by additives of transition metals compounds and phosphorus-containing additives (PCA).

The first of the mentioned facts is explained by dependence of polymer oxidation kinetics on molecular mobility of macrochain. Combination of pyromellitimide fragment and aliphatic chain with even number of methylene groups provides for forming rigid high-crystalline PAI-12 structure {$(CH_2)_{12}$ methylene chain length}. The physical structure of the closest “uneven” homologues {$(CH_2)_{11}$ and $(CH_2)_{13}$ in PAI-11 and PAI-13, respectively} is less organized (wide bands NMR data), and consequently, these polymers are oxidized faster in the solid phase.

The rest facts should be explained:

- PDI extraction requires synchronization of the oxidative process in neighboring methylene chains, separated by a polymellitimide fragment. Random chain initiation and transmission would cause occurrence of a series of oligomers in products of solid-phase oxidation;

- H_2 release and cross-linking macrochains during oxidation at relatively low temperature is untypical for methylene chains of aliphatic polymers, but this phenomenon accompanies oxidative aging of fully aromatic PI and PPQ;
- Transition metals compounds stabilize PAI at both low and high temperatures, but sharply accelerate thermooxidation of methylene chains, for example, PE.

Search for an alternative or an addition to the radical-chain scheme indicated new properties of PAI. Judging by shear in electron spectra of the model compound, didodecylpyromellit diimide, at solution blowing through by argon first and then by oxygen, polymellitimide structure forms a molecular complex with O_2. Preliminary PAI treatment by I_2 possessing a strong complex forming effect leads to a significant decrease of the 1st exothermal DTA peak characterizing the initial oxidation phase at dynamic heating. Moreover, study of magnetic properties indicated PAI paramagnetic properties. The polymer displays ESR signal – the singlet with $q_{av} = 2.183$. Measurements of magnetic susceptibility by the separate Faraday method shows that PAI possesses the magnetic moment $\mu_{eff} = 0.77 – 0.94$ M.B. (297 K). Temperature dependence of μ_{eff} in the range of negative temperatures is typical of cluster compounds. The value of exchange parameter $\zeta_{i,j} = 226$ cm^{-1} points to high conjugation degree along the polymeric chain or between chains. As chemical structure of PAI with a long chain of σ-bonds does not provide for the conjugation system along the chain, the unique possibility of its realization is formation of intra- and intermolecular complexes between pyromellitimide fragments. $CuSO_4$, the effective PAI thermostabilizer, possesses $\mu_{eff} = 1.86$ M.B. The additivity rule of total molar magnetic susceptibilities for mechanical mixtures of PAI and $CuSO_4$ is not obeyed both at room temperature and after thermal treatment of the mixture at 320°. Thermal treatment in vacuum, in the air or in the inert gas medium leads to a significant increase of magnetic susceptibility and efficient magnetic moment of PAI and PAI mixture with $CuSO_4$, but the effect is much higher at thermal treatment in the air. Studies of magnetic properties point to the interaction of macromolecule and additive, this interaction being obviously realized by pyromellitimide fragment.

To "include" pyromellitimide fragments into oxidation independently of methylene chains, the initiation act of thermooxidation is suggested via formation of O_2 molecular complexes with π-system and/or thermal transmission of PAI conjugated system (cluster) into electron-excited state, from which radical reactions and transmission of H atoms are initiated, in particular, cross-linking of aromatic rings with H_2 release. Adding initiation of thermooxidation radical-chain scheme by these acts gives a possibility to explain regularities of PAI thermooxidation. For example, efficiency of transition metals compounds as thermostabilizers can be associated with formation of the additive–π-system complex considering this act as the one competitive to occurrence of a molecular complex with O_2.

3. Thermooxidative Degradation of Polysulfones, Polyesterketones, Liquid-Crystal Copolyesters [21, 30 – 39]

Mass losses at dynamic heating of polysulfone based on bisphenol A and dichlordiphenylsulfone (PSF) in the absence or in the presence of O_2 is characterized on the initial stage by rate constants equal $2.4 \cdot 10^{13} \cdot \exp(-190/RT)$ and $4.3 \cdot 10^{7} \cdot \exp(-92/RT)$ min^{-1}, respectively, i.e. in the absence of O_2 decomposition is sharply accelerated. O_2 absorption and release of CO_2, H_2O and CH_2O accompany PSF heating at 300 – 350°. By the shape of kinetic curves, PSF oxidation is similar to other above-described PHA: two-stage kinetics with the fast initial stage lasting 2 – 4 hours and the second slow stage with total O_2 absorption depending on temperature of, for example, 1 – 2 mol/base-mol during next 10 hours and with effective activation energies of 72 and 11 kJ/mol on both stages, respectively. As oxygen partial pressure reduces from 400 to 30 kPa, the oxidation kinetics (by O_2 absorption and extraction of appropriate products) changes from σ- to s-shape. Thermooxidation in melt leads to gel formation and PSF yellowing. These changes are symbathic to O_2 absorption and depend on its pressure. At 300 – 350° the gel formation point corresponds to 1.5 – 3 hour exposure. However, the raise of all MMD moments, especially M_z, is observed since the very beginning of thermooxidation. These changes are realized by branching PSF macromolecular, the branching frequency (λ) increasing linearly with exposure, finally reaching the limit value $\lambda = 0.25 - 0.36$ per base-mole near the gel formation point. The calculation was performed for the long-chain model that allowed determination of the intercross-link point distance as ≅ 4 – 5 thousands. In the absence of O_2 at 320°, an insignificant branching ($\lambda = 0.0055$) is observed only 10 hours after, whereas the analogous process in the air gives $\lambda = 0.008 - 0.015$. The effect of shear loads on branching under the conditions of PSF thermooxidation is rather low. For example, in the test on rheoviscosimetrs device, λ increases by 20% only with 10-fold load. ^{13}C NMR-spectra of PSF, oxidized in melt, display new signals of substitutes and non-substituted aromatic carbon atoms with chemical shears at 140 – 150 and 110 – 120 m.p. IR-spectroscopy data also confirm new substitutions of aromatic rings and expense of the main p-phenylene structure. One has suggested that branchings occur from aromatic C atoms with branching points of the following types:

This enables to introduce a concept of polar effects in homological reactions, known in kinetics of radical polymerization and homolytic arylation of p-substituted benzenes. Combination of heterogroups with different electronegativity in the PSF structure – strong electron-acceptor sulfonyl (the Hammet constant $\sigma_n = +0.7$) and electron-donor ether ($\sigma_n = -0.32$) groups, as well as weak electron-donor isopropylydene group ($\sigma_n = -0.197$), causes alternation of electron density on aromatic C atoms, which is naturally displayed in values of chemical shears of ^{13}C NMR:

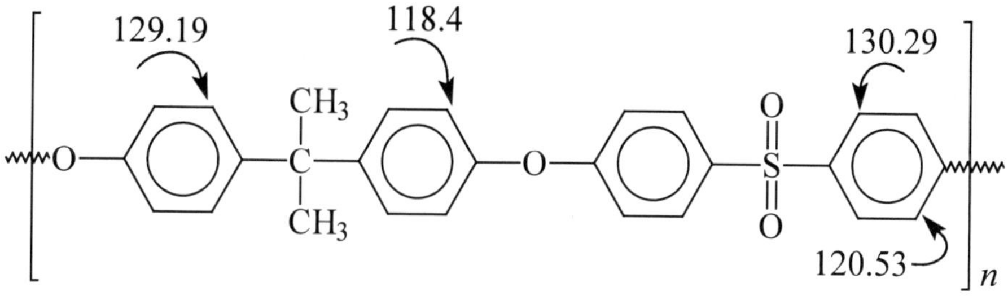

As chemical shears of neighboring ring atoms C differ by ~10 m.p., the difference in π-electron density reaches 0.055 e at the average proportionality coefficient of 180 m.p. per electron. Aromatic radicals also display polar effects of p-substituents. PSF structure, as well as the one of polyestersulfones (PSF purely aromatic homologues, the ones based on dioxydiphenylsulfone and dichlordiphenylsulfone and/or bis(4-chlorphenylsulfonyl) biphenol), is conductive to radical attachment decreasing activation barrier of the reaction at the sacrifice of the Coulomb interaction. The possibility of this qualitative approach is confirmed by weak, comparing with PSF and PES, structuring of polycarbonate (PC) under equal conditions – the polyester, produced similar to PSF on the base of bisphenol A. Judging by ^{13}C NMR, aromatic p-substitution in PC induces weaker alternation of π-electron density by ring C atoms. Homolytic arylation substitutes active radicals by hexadienyl or arylhexadienyl ones, relatively stable due to polar effects of substituents. Calculation of the bond energy shows that C-CH_3 in the isopropylidene fragment is a relatively weak bond in PSF and equals ≅220 kJ/mol. By all fragments of the structure, the above-mentioned aromatic PSF homologues are characterized by bond strengths, leveled as ≅330 – 450 kJ/mol, and possess higher thermoresistance in the absence of oxygen, respectively. At thermooxidation, O_2 greatly contributes to degradation. In the case of classical thermooxidation mechanism, O_2 attack is aimed at aliphatic C-H-bonds under the condition of interlocking end phenolic groups. In response to PSF oxidation, only about 20% of absorbed O_2 transit into volatile products: CO_2, H_2O and CH_2O. The main part of O_2 remains in PSF in the form of polymeric oxidation products. Aliphatic, C=O-, and aromatic, –OH- and C=O-, groups, accumulated simultaneously with expending CH_3, are found in the oxidized polymer. Accumulation of aromatic oxygen-containing groups in PSF and PES is already noted on the initial stage of the process, corresponded to the sigmoidal part of the kinetic curve of O_2 absorption. For example, about 60 – 80% of CH_3-groups remain after thermooxidation at 320° during 15 – 20 hours. Hence, the universally accepted model of separate oxidation of aliphatic fragments first, and then aromatic ones is excluded. High-temperature oxidation touches upon and develops in aromatic and isopropylidene fragments of the PSF macrochain simultaneously. Similarity of oxidation gross-kinetics of PSF and PES, as well as of polyesterketones (PEK) points on participation of aromatic rings in oxidation already on initial stages. Difference in thermoresistance is determined by microchain rigidity. Thermoresistance decreases (according to dynamic TGA data in the air) in the sequence PEK > polydiphenylesterketone > PES > polyesterestersulfone > PSF. Therewith, initially linear macromolecule gradually transforms to a structured macroformation, partly losing the initial structure, but with new elements developing a system of conjugation blocks, which induces polymers yellowing:

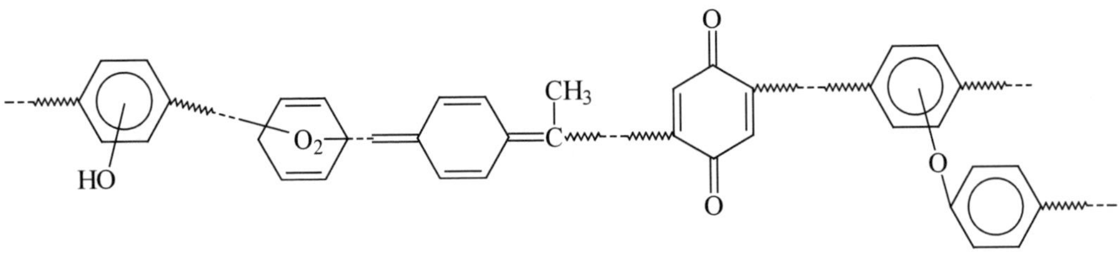

Beside the variant of typical radical-chain scheme development with future oxidation transfer to aromatic rings at H atom detachment in the chain propagation reaction, the possibility of direct attaching oxygen to aromatic rings with forming endoperoxides of peroxyradicals is discussed for the mechanism of high-temperature oxidation. Similar to previously considered PHA, this reaction is probably accompanied by forming a molecular complex of π-system with O_2 and/or thermal transfer of the conjugated system to the electron-excited state. This supposition seems quite real, keeping in mind that the red limit of electron absorption spectra of PES and PSF locates at 400 – 500 nm. This also explains high resistance of polymer containing $-SO_2$-group in the structure to thermooxidation. A strong electron-acceptor bridge $-SO_2$- inhibits formation of molecular complexes with O_2. Additional argument to confirm the mechanism suggested is the highest stabilizing efficiency of phosphorus-containing additives, taking into account the complex forming function of phosphorus.

In recent 10 – 15 years liquid-crystal aromatic polyesters (LCP) became of special interest, which above all was associated with the discovery of their ability to self-reinforcement in melt. This enables to prepare high accuracy and size stable articles with low coefficients of thermal shrinkage from them.

Special interest to these polymers is stipulated by intensive progress in computer technology. In recent 5 – 10 years, a great list of publications concerning structure and properties of LCP was presented. However, information on thermoresistance and specific features of LCP degradative behavior is short. Authors have studied degradative transformations of LPC, based on p-oxybenzoic acid, dioxydiphenyl and terephthalic and isophthalic acids (TPA and IPA). Therewith, TPA/IPA concentration varied in the range of 100/0 – 0/100.

Judging by kinetics of oxygen absorption at 350° (processing temperature) and data of dynamic TGA/DTA, thermooxidative stability of LCP with various concentrations of TPA/IPA decreases in the sequence: IPA-0 > IPA-25 ≥ IPA-50 > IPA-75 > IPA-100.

In the absence of oxygen, LCP decomposition proceeds in one stage with formation of a significant (up to 40 wt.% at 700°) coke residue. Two endothermal heat effects were observed on DTA curves: low peak in the melting area and quite intensive one at polymer degradation. Value of the heat effect, associated with melting, is $\Delta H = 1 - 2.5$ kJ/mol. In ref. [38] the authors, who studied 4-hydroxybenzoic and 2,6-hydroxynaphthoic acids copolymer by the differential scanning calorimetry (DSC) method, believe that the heat effect of about 1 kJ/mol relates not to the real melting, but to a change of ordering degree at transition from a crystal to mesophase.

LCP thermoresistance in the air is much lower (~ by 25 – 30°) than in argon. According to dynamic TGA data in the air, mass losses of studied LCPs are observed within the range of

350 – 800°. Decomposition proceeds in two stages: the first stage at 350 – 550° is accompanied by mass losses of up to 40 wt.%, and the second slower one in the temperature range of 550 – 800° proceeds practically to complete degradation of polymers. Coke content at 750 – 800° is 3 – 10 wt.%. Judging by DTA data, the stages of LCP degradation are accompanied by exothermal heat effects. In tests in the air, weak endothermal heat effect is absent in the "melting" range, probably because of overlapping by exothermal effects accompanying degradative reactions. Increase of IPA concentration leads to a shear of the melting range to the area of low temperatures and reduction of LCP thermoresistance.

Study of LCP phase transitions by the X-ray structuring analysis (RSA) in the temperature range of 20 – 400° showed that similar changes proceed in all LCPs. Annealing at 300° causes an insignificant increase of the main crystal reflexive action. The ability of studied materials to transit into the so-called "liquid-crystal state" displays their behavior at processing temperature. In all polymers, a jump-like viscosity decrease is observed at softening temperature (depending on the structure, it falls within the range of 300 – 400°). Therewith, very strong threads are formed from the melt. This effect is explained [39] by cooperative orientation of large axes of macromolecules along the flow direction (viscosity anisotropy), which is realized only in LCP.

It is common knowledge [7, 21, 39] that thermoresistance of polymers depends upon several factors: structure, molecular-weight characteristics, content of macrochain defects, labile end groups (hydroxyl one, in the present case), low-molecular organic admixtures (unreacted monomers residues) or inorganic ones (admixtures of metal ions appearing in a polymer from the raw material and equipment) in macromolecules.

The composition of inorganic admixtures in monomers and LCPs was studied by the ICP method. Nonequivalent influence of some metal admixtures on thermoresistance was shown by the TGA method on the example of LCP KI-75. Introducing inorganic salts of an appropriate metal into the polymer increased concentration of admixtures. The results obtained indicated different influence of metal admixtures on thermooxidative resistance of LCP. Aluminum and metals from the alkaline sequence, such as Ca, Na, K, do not practically affect thermoresistance in the studied range of concentrations. Fe, and Ca and Ni induce a negative effect in concentration up to $10^{-3} – 10^{-4}$ wt.% significantly increase thermoresistance of LCP.

Composition and concentration of organic admixtures in LCP were studied by the mass-spectroscopy method. Admixtures of phenol and dioxydiphenyl (m/e 94 and 186) – the hydrolysis product of the initial monomer (dioxydiphenyl diacetate), p-oxybenzoatedioxydiphenyl and oxydiphenyl acetate were identified. Total amount of organic admixtures in different samples did not exceed $(1.0 – 2.0)\cdot 10^{-2}$ wt.% that had no effect on thermoresistance of polymers.

Oxygen absorption kinetics is the two-stage process consisting of the first fast stage (2 – 3 hours long) with absorption of more than 1 mole of O_2 per molecular unit and the second slow one with absorption of 0.2 mol/base-mol O_2 during next 7 – 8 hours of thermooxidation. Analogous dependencies are displayed by CO_2 release kinetics – the main gas product of studied LCPs degradation. Dioxydiphenol and p-oxybenzoic acid derivatives are present in the composition of heavy high-boiling degradation products, identified by NMR and MS methods: diphenylbenzoate, oxyphenylbenzoate, etc. The composition of the above-mentioned products is the same for all studied LCPs and differs only by ratio of the components. Analysis of the kinetic curves shows that thermooxidation rate is much lower at lower temperatures of 300° and 320°, near the melting point or at melting, than at 350°, when, judging by RSA data, the polymer completely transits into the isotropic melt. Beside the main CO_2 gas product, hydrogen (on early stages of oxidation at exposure of 0.5 – 1.0 hour) and water (exposure longer than 4 hours) were identified at LCP thermooxidation in the studied temperature-time limits. Dynamics of the element composition change shows that hydrogen concentration decreases and carbon concentration increases in the sample containing IPA-75 during thermooxidation, i.e. a graphite-like structure is formed. This process proceeds intensively at 350°. IR-spectroscopy data show that initial changes happen in the area of absorption of ester aromatic fragments: intensity of absorption bands decreases at $\nu_{C=O}$ = 1740 cm^{-1}, ν_{C-O} = 1270 and 1160 cm^{-1}, and $\nu_{C=C}$ = 1600 and 1500 cm^{-1}, $\delta_{C=C}$ = 720 cm^{-1}. Absorption bands of ether bonds, ν_{C-O-C} = 1080 cm^{-1}, and aromatic structures are the only ones remained at the maximal exposure (thermooxidation at 350° during 10 hours). Therewith, the spectrum noise decreases, which is also associated with formation of intermolecular cross-links. At 350°, a great amount of oligomers, precipitated in the ampoule neighboring the reaction zone at ≅150°, is formed in melt. Structure and ratio of oligomers with p-oxybenzoate dioxydiphenyl (product a), oxydiphenyl (product b), p-oxybenzoic acid di- and trimers (product c) end groups in the mixture of oligomeric degradation products are identified with the help of ^{13}C NMR.

The amount of oligomers in the mixture was successfully determined by the ratio of signal squares with appropriate chemical shears of 11.5 and 115.82 ppm (product a), 115.43 ppm (product b) and 120.72 ppm (product c): product a – 15 mol.%, product b – 31 mol.%, product c - ≅10 – 12 mol.%.

The fact of detecting neighboring p-oxybenzoic acid units in the composition of thermooxidation products allows a supposition that, beside copolycondensation, homopolycondensation reactions of p-acetoxybenzoic acid proceed also. The output of free p-oxybenzoic acid is 3 – 4-fold higher, than its output in the linked state shaped as end groups in oligomers. Apparently, free p-oxybenzoic acid is formed as the result of thermal reactions at decomposition of labile bonds in structural p-oxybenzoic blocks.

Analysis of kinetics and composition of LCP degradation products in the processing temperature range cleared up some general features in degradative behavior of thermostable PHA [41]: structure graphitization, H_2 release, thermooxidation resistance increase as additives of transition metals and PCA are introduced.

4. Thermooxidative Degradation of Polyesterimides, Polyamidoimides, Aliphatic-Aromatic Polyamides [7, 21, 40 – 43]

Water is the main product (~0.3 wt.%), extracted from polyesterimide (PEI) based on dianhydride A (bisphenol A diaminodiphenolic ester) and m-phenylene diamine by thermooxidation in the temperature range of 320 – 340° at exposure up to 1 hours. The rest of the products give ~0.1 wt.%. The following compounds were detected: CO, CO_2, residues and transformation products of initial monomers, for example, bisphenol A. Increase of exposure duration and/or temperature up to 420° (initiation of intensive degradation by TGA) leads to deeper transformations of the polymeric structure and, consequently, to formation of greater amounts of products with different compositions. IR-spectroscopy data show that short-term thermooxidative effect causes expense of isopropylidene structure and p-substitution of aromatic rings at stability of phthalimide cycles. Gel-penetrating chromatography (GPC) data indicate polymer structuring intensified with temperature or O_2 content in the system. Temperature raise up to 420° causes M_z increase by more than 30 thousands, increase of all moments of molecular mass and polydispersion from 2.3 to 3, respectively. The changes mentioned manifest themselves as a reduction of light transmittance of solutions and PEI yellowing, and at deeper changes – as a formation of insoluble gel-fraction, which content increases symbathly to O_2 absorption. The following compounds were identified in heavy (oligomeric), volatile (gaseous), solid and liquid products of PEI thermooxidation with the help of IR-spectroscopy, NMR, mass spectrometry (MS), highly efficient liquid chromatography (HELC) and gas chromatography (GC): H_2, CO_2, CO, CH_2O, CH_3OH, H_2O, phenol and derivatives, bisphenol A, aromatic and ester-aromatic compounds with methyl, ethyl and isopropylidene groups. Judging by mass losses and melt viscosity stability, PCAs display the highest thermooxidative efficiency in PEI.

Judging by TGA data, thermooxidative resistance of polyamidoimide (PAIM), based on trimellitic acid anhydrochloranhydride and dianhydride A, is by ~10 – 20% lower than that of purely aromatic PAIM (based on trimellitic dianhydride and m-phenylene diamine). Similar regularity is observed in the study of thermooxidation in melt (higher thermoresistance for rigid-aromatic PAIM structure). Usual sigmoidal kinetics of O_2 absorption and CO_2 and CO release with fast first and slow second stages is typical of both polymers. Analysis of thermooxidative effect efficiency of numerous additives of various classes (amines, phenols, benzophenols, hydrazines, etc.) indicated the unambiguous advantage of PCA. As PCA is introduced, polymer yellowing at exposure in melt is reduced, mass losses are decelerated and melt viscosity is stabilized.

Thermooxidation of polyphthalamides (PPA-1 and PPA-2) displays the same regularities (kinetics of mass losses, oxygen absorption and release of the main volatile products, composition of volatile and heavy products relative to the polymer structure, oxidation inhibiting by adding PCA and transition metal compounds, etc.), which are described in detail for PAI, PPQ, PBO, PEI, PI, PAIM, PSP, PES and LCP. That is why, not leaning on experimental results, let us discuss in more details the degradation mechanism on the example of these two polymers.

Structure of polyphthalamides PPPA-1 and PPA-2 represents alternation of aliphatic and aromatic sequences.

PPA-1

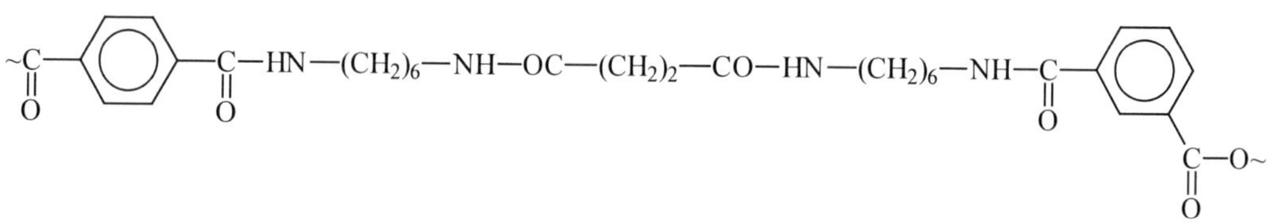

PPA-2

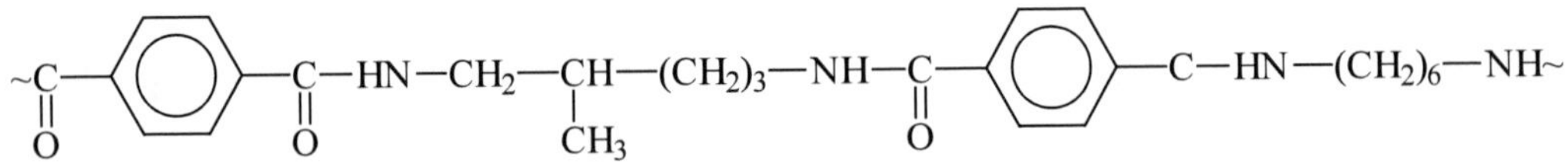

As thermooxidation develops statistically along the macromolecule, amide of terephthalic acid (TPA) formation (attack 1 – 2) is of the same probability as other fractures of the macrochain (attack 3 – 4) (scheme 2),

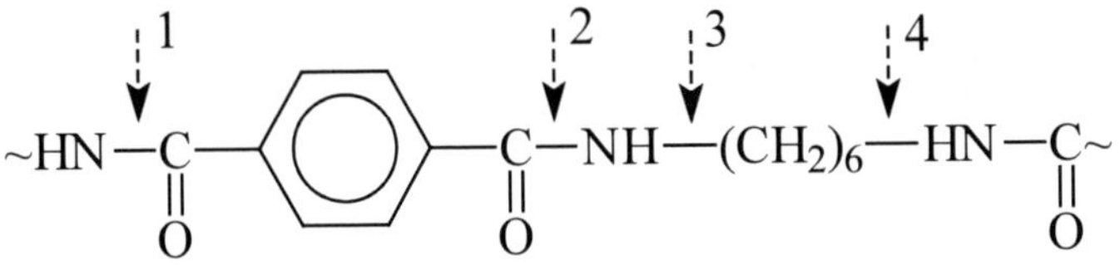

TPA amide is formed by thermooxidation in melt and at low-temperature solid-phase oxidation. Light yellow crystals of TPA amide and TPA mixture occur and accumulate on the surface of mold samples during accelerated heat aging at temperatures of 150 – 200°C. TPA is a typical product of hydrolysis by macrochain ends. To our point of view, TPA amide product is of greater interest. Recall that analogous products (relative to polymer structure) were already observed in investigation of thermooxidative transformations of other PHA. For example, pyromellite diimide,

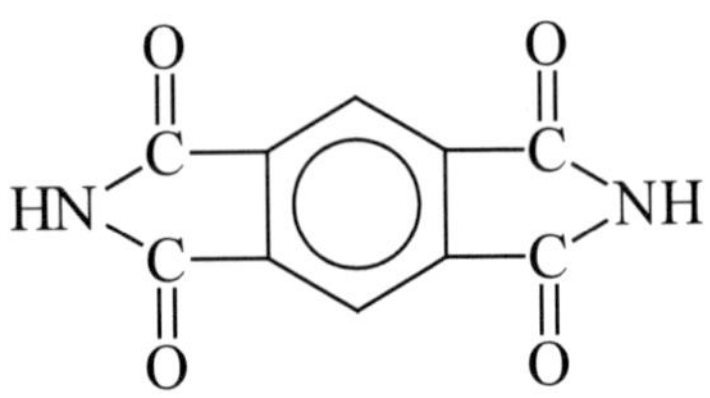

was identified in thermooxidation of PAI [7, 21] and classical polyimide "Capton", and 2,2'-(1,4-phenylene)-bis-(phenylpyrazine),

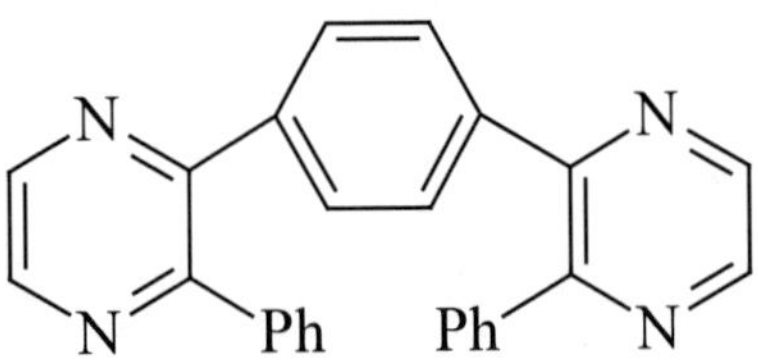

is formed at poly(phenylquinoxaline) aging [7].

To our point of view, a specific feature of thermooxidation of aromatic and aliphatic-aromatic heterochain polymers is displayed in TPA amide formed during PPA thermooxidation. Unfortunately, this feature is insufficiently discussed in the literature yet. Three hypotheses were put forward in discussion of PDI formation at PAI thermooxidation:

1. Sequential burning of (oxidation) of aliphatic fragments from both sided of the imide cycle.
2. Formation of a molecular complex with oxygen [43].
3. Macromolecule transition into electron-excited state [11].

The first hypothesis is nonviable, because neither direct, nor indirect proofs were found for it. The authors [21] confirmed the possibility of molecular complex with oxygen existence indirectly – on model compounds, and transition of macromolecules into the electron-excited state as the act of radical-chain process initiation was proved by quantum-chemical calculations of models and PHA oligomers. Probably, formation of TPA amide, in products of PPA thermooxidation, as well as PDI in PAI and PI, can be explained by existence of two latter hypotheses.

Introduction of additives is the common method of studying the mechanism of chemical reactions. Phenolic antioxidants, inhibiting oxidation in reactions with peroxyradicals RO_2, in concentration of 10^{-2} – 10^{-3} mol/kg decelerate O_2 absorption by PPA-1 and PPA-2 at 150 – 250° (solid-phase oxidation). Kinetics of O_2 absorption by PPA is characterized by absence of the induction period, e.g. critical concentration of antioxidant cannot be found. Concentration dependence of the antioxidant efficiency, determined from the relation of initial rate constants of O_2 absorption by unstabilized and stabilized PPA-2, displays a clear optimum at concentration of 0.008 mol/kg. Although at phenols introduction antioxidant effects are not sufficiently high, probably due to high temperature of PPA processing, the fact of decelerating thermooxidation by itself (for example, at phenolic antioxidant introduction in concentration of 0.2%, O_2 absorption by PPA-1 reduces by 10 – 20%) confirms once again contribution of the radical-chain process into PPA thermooxidative degradation.

Some features of PPA thermooxidation do not fit in the traditional scheme. The supposition about molecular complexes with oxygen lead to PPA analogies with PAI, PI, PPQ, and other PHA, thermooxidation of which are decelerated by additives Cu^{2+} [11] and PCA. For example, addition of anilidophosphoric acid diphenyl ether and $CuSO_4$ was found the most effective in PI and PAI [7, 21]. The mentioned compounds as additives decelerate O_2 absorption in PPA-2 at the solid-phase oxidation noticeably more effectively, than phenolic antioxidants. Efficiency of the additives is also noticeable at high temperatures, at which phenols are inefficient.

The concentration dependence of Cu^{2+} adding efficiency passes through an extreme, as in the case of aliphatic PA. Even micro concentrations of Cu^{2+} are quite effective for stabilizing PPA. It is common knowledge that introduction of Cu^{2+} salts into PE destabilizes methylene chain at both low (solid-phase oxidation) and high (processing) temperatures. Comparison of Cu^{2+} additive effect, for example, in PPA, PAI, PSP and PE, points to a significant role of aromatic fragment in the oxidation.

Flexible methylene chains, as any hydrocarbonic sequences, are the fragments most vulnerable for thermooxidation. Rigidity of the physical structure, the limitedness of molecular motions, in PAI is defined by pyromellitimide fragment, and in PPA-1 and PPA-2 – terephthalic one. The presence of IPA in PPA-1 also imparts a definite flexibility to the macromolecule (similar to the above-considered LCP). To put it differently, reducing molecular mobility of the macrochain, aromatic fragments stimulate physical mechanisms of methylene branchings protections from thermooxidation.

From positions of chemistry, if physical limits are neglected, the aromatic fragment increases reactivity of border methylene groups, which are more active in reaction with O_2. Oxidation restrained in the solid phase develops violently in the polymer melt, touching upon aliphatic parts first and aromatic fragments in lesser degree.

Some specific features of PPA-1 and PPA-2 thermooxidation are displayed at low temperatures on the background of the typical radical-chain process. Antioxidative effect of phenolic antioxidants is much weaker, than that of copper-containing additives and phosphates. Oxidation causes an ejection of TPA amide from macromolecule and an intensive cross-linking of macrochains. One may suppose participation of aromatic fragments in degradation, therewith, not only by α-bonds activation in the neighboring flexible chain. Although direct proofs are absent yet, the authors suggest that, beside classic initiation, for example, of the type

$$RH + O_2 \rightarrow R^{\bullet} + RO_2^{\bullet},$$

where RH is the methylene chain, thermooxidation is initiated with participation of aromatic phthalamide fragments. This can be formation of a molecular complex of one or several O_2 molecules with π-system of the fragment, or thermal transformation of the fragment into the triplet state with the help of O_2 (similar to photooxidation). Probability or improbability of these facts is discussible. No doubt that "for these polymer the classical radical-chain scheme of oxidation needs a significant supplements". The following initiation reactions are possible:

1. Classic reaction:

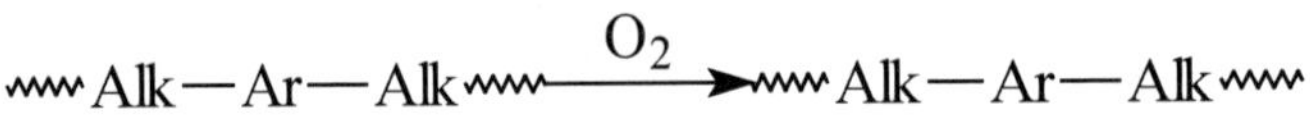

2. Formation and degradation of molecular complex with O_2:

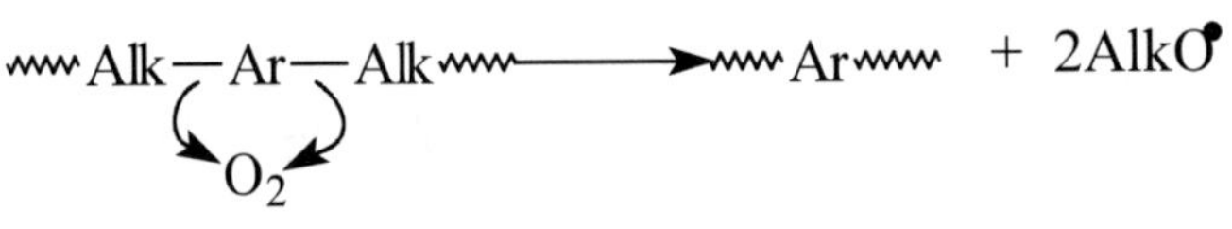

3. Transformation into the electron-excited state:

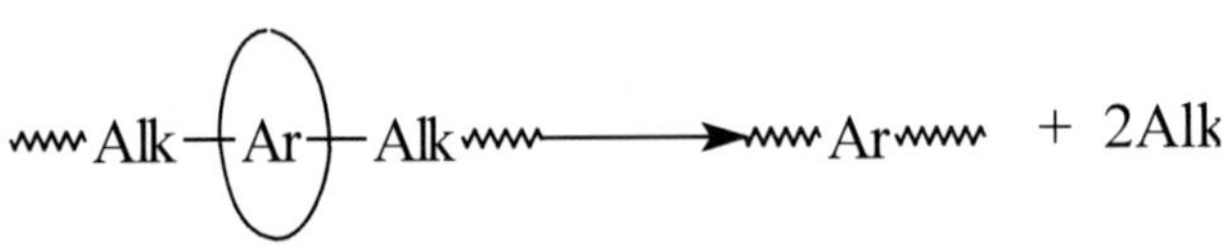

The succeeding stages fit in with the radical-chain scheme and elementary acts of the chain transmission by end and middle alkyl radicals lead to formation of all detected volatile, low-volatile and polymeric products of degradation.

In the applied aspect, the suggested scheme of PPA thermooxidative degradation, which to the authors' point of view is applicable to all above-mentioned PHA, indicates two directions of antioxidative stabilization – inhibition of the radical chain process and "destabilizing" function of the aromatic fragment. Therewith, the applied task is subdivided into two directions: antioxidative stabilization during processing (temperature range of 300 – 350°, short-term) and stabilization during exploitation (working temperatures within the range of 150 – 200°). The authors suppose that the account for features of thermooxidation gives grounds for optimization of thermostabilizing compounding of all "life stages" of thermoresistant PHA by combining antioxidants with compounds of alternating valence metals and phosphorus-containing substances.

REFERENCES

[1] Blumenfeld A.B., Vdovina A.L., Voloshchuk K.A., Zabel'nikov N.S., and Tseitlin G.M., Thermal aging of thermoresistant polymer poly(phenylquinoxaline), *Proc. XV Colloquium of Dunarea Countries on the Problems of Natural and Artificial Aging of Polymers*, September 27 – 30, 1982, Moscow, p. 4142. (Rus)

[2] Vdovina A.L., Blumenfeld A.B., Zabel'nikov N.S., and Annenkova N.G., On thermoresistance of phenylquinoxaline copolymers, *Vysokomol. Soeding.*, 1984, vol. **26B**, No. 3, pp. 223 – 227. (Rus)

[3] Vdovina A.L., Blumenfeld A.B., Annenkova N.G., Pleshkova A.P., Zabel'nicov N.S., Kachevskii O.V., and Ryaboi V.M., Regularities of polypyromellitimide and poly(phenylquinoxaline) thermooxidative transformations, *Proc. XXII All-Union Conference on High-molecular Compounds*, Section and stands thesis, October 15 – 19, 1985, Alma-Ata. (Rus)

[4] Vdovina A.L., Ryaboi V.M., and Blumenfeld A.B., Quantum-chemical criterion of thermooxidative resistance of poly(heteroarylenes), *Vysokomol. Soedin.*, 1985, vol. **B27**, No. 11, pp. 808 – 811. (Rus)

[5] Blumenfeld A.B., Vdovina A.L., Pleshkova A.P., and Annenkova N.G., About products of polypyromellitimide and poly(phenylquinoxaline) therooxidation, *Vysokomol. Soedin.*, 1986, vol. **B28**, No. 7, pp. 500 – 503. (Rus)

[6] Vdovina A.L., *Thermal transformations and stabilization of polypyromettilimide, poly(phenylquinoxaline) and co-poly(imidophenylquinoxalines)*, Candidate Diss. Thesis on chemical sciences, Moscow, G.S. Petrov RIPM, 1988. (Rus)

[7] Tseitlin G.M., Voloshchuk K.A., and Blumenfeld A.B., Thermal degradation of poly(phenylquinoxaline), *Vysokomol. Soedin.*, 1989, vol. **A31**, No. 2, pp. 295 – 297. (Rus)

[8] Blumenfeld A.B., Vdovina A.L., and Annenkova N.G., Thermal behavior and heat stabilization of polyimide and related polymers, In: *Polyimides and Other High-temperature Polymers*, Eds. M.J.M. Abadie, B. Sillon, Elsevier Publ., 1991, pp. 183 – 189.

[9] Buchachenko A.L., Yasina L.L., Vdovina A.L., and Blumenfeld A.B., New nuclear spin selective reaction of molecular oxygen, *Izvestiya AN, Ser. Khim.*, 1994, No. 8, pp. 1402 – 1404. (Rus)

[10] Blumenfeld A.B., Levantovskaya I.I., and Annenkova N.G., Thermal stability and stabilization of heterochain polymers, *Itogi Nauki i Tekhniki*, 1985, vol. **20**, VINITI, Moscow, pp. 143 – 211. (Rus)

[11] Annenkova N.G., Blumenfeld A.B., Vavilova L.A., Pinaeva N.K., Savina M.E. and Chernova A.G., Thermostability and stabilization of fatty-aromatic polyimide, *International Subject Collection "Polycondensational Processes and Polymers"*, 1985, Nal'chik, pp. 16 – 24. (Rus)

[12] Abramova I.M., Azriel A.E., Vavilova L.I., Vatagina V.A., Zezina A.G., Zyisk K.Yu., Kazaryan L.G., Pinaeva N.K., Savina M.E., Sade A.G., and Chernova A.G., Influence of aging conditions on structure and properties of polyalkanimide, AI-1G trademark, In: *Thermoresistant Materials*, 1985, Moscow, NIITEKhim, pp. 162 – 167. (Rus)

[13] Tuichiev Sh., Kuznetsova A.M., and Mukhametdieva A.M., Temperature dependencies of molecular and supermolecular structure of polyalkanimide, *Vysokomol. Soedin.*, 1987, vol. **A29**, No. 8, pp. 1756 – 1760. (Rus)

[14] Kalugina E.V., Blumenfeld A.B., Annenkova N.G., Gvozdev D.V., Savina M.E., and Pinaeva N.K., Thermooxidative stability of polyalkanimide, *Plastmassy*, 1988, No. 7, pp. 26 – 28. (Rus)

[15] Kalugina E.V., Blumenfeld A.B., Annenkova N.G., and Arshava B.M., Combustibility and degradation of polyalkanimide, *Proc. VI All-Union Conference "Polymer combustion and creation of limitedly combustible materials"*, 1988, Sugdal', p. 64. (Rus)

[16] Kazaryan L.G., Azriel A.E., Vasil'ev V.A., Annenkova N.G., Pinaeva N.K., and Chernova A.G., Structure and properties of polyalkanimide, *Vysokomol. Soedin.*, 1988, vol. **A30**, No. 3, pp. 644 – 647. (Rus)

[17] Kalugina E.V., Blumenfeld A.B., Annenkova N.G., Savina M.E., and Arshava B.M., Degradation and stabilization of polyalkanimide, Tes. Rep. IX All-Union Conference "Synthesis and Efficiency Research of Chemicals for Polymeric Materials", 1990, Tambov, p. 49. (Rus)

[18] Sekacheva I.V., Solynkin I.N., Pinaeva N.K., Koshel' G.N., Kyukova G.G., Kozlov O.S., and Smirnova O.M., Synthesis and identification of salts based on pyromellitimidic acid and 1,12-dodecamethylenediamide, *Yarosl' Polytechnical Institute Coll.*, 1990. (Rus)

[19] Kalugina E.V., Blumenfeld A.B., Annenkova N.G., Arshava B.M., Kotov Yu.I., Savina M.E., and Pleshkova A.G., Thermostability and thermal transformations of polyalkanimide, *Plastmassy*, 1991, No. 7, pp. 48 – 51. (Rus)

[20] Kalugina E.V., Thermal transformations and stabilization of some thermoresistant heterochain polymers, *Thes. Candidate Diss. on Chemical Sciences*, 1992, Moscow, G.S. Petrov RIPM. (Rus)

[21] Kalugina E.V., Blumenfeld A.B., Annenkova N.G., and Volkov M.A., Stabilization of aliphatic-aromatic polyimide, *Plastmassy*, 1993, No. 1, pp. 30 – 32. (Rus)

[22] Blumenfeld A.B., Kalugina E.V., Annenkova N.G., Volkov M.A., and Savina M.E., Heat effects in aliphatic-aromatic polyimides, *Plastmassy*, 1993, No. 1, pp. 24 – 27. (Rus)

[23] Kalugina E.V., Blumenfeld A.B., Annenkova N.G., Volkov M.A., Savina M.E., Rafieva S.R., Markova E.I., and Janibekov N.F., Polyalkanimide thermostability increase with the help of special additives, *Plastmassy*, 1993, No. 3, pp. 30 – 32. (Rus)

[24] Kalugina E.V., Blumenfeld A.B., Savina M.E., and Novotortsev V.M., Thermostabilization of aliphatic-aromatic polyimide at processing and long-term heating aging, *Plastmassy*, 1996, No. 1, pp. 14 - 15. (Rus)

[25] Savina M.E., Blumenfeld A.B., Chernova A.G., and Kalugina E.V., *Russ. Patent No. 2,028,337, Polymer composition*, 1992. (Rus)

[26] Savina M.E., Chernova A.G., Kalugina E.V., and Astakhov P.A., *Russ. Patent No. 2,069,670, Production method of fatty-aromatic polyimides*, 1993. (Rus)

[27] Blumenfeld A.B., Kalugina E.V., Bolotina L.M., and Savina M.E., The problem of thermoresistance of high-temperature engineering thermoplasts, *Plastmassy*, 1993, No. 2, pp. 21 – 24. (Rus)

[28] Blumenfeld A.B., Kalugina E.V., and Zaikov G.E., Structure. Impurities. Thermal stability and thermal stabilization of the engineering thermostable resins of polyimide and polysulfone types, *Polymer Yearbook*, Ed. Richard A. Pethrick, 2000, vol. **17**, Harwood Academic Publ., pp. 275 – 285. (Rus)

[29] Narkon L.A., Research of additives, end groups and molecular-weight characteristics effect on polysulfone thermoresistance, Thes. Candidate Diss. on Chemical Sciences, 1984, Moscow, G.S. Petrov RIPM. (Rus)

[30] Levantovskaya I.I., Narkon A.L., Ershov O.V., Gur'yanova V.V., Reiburd L.I., Radetskaya M.P., Bolotina L.M., Blumenfeld A.B., Pavlov A.V., and Motorina M.P., Research of some regularities in polysulfone high-temperature oxidation, *Vysokomol. Soedin.*, 1985, vol. **A27**, No. 2, pp. 362 – 368. (Rus)

[31] Pavlov A.V., Gurjanova V.V., Ershov O.V., Kalugina E.V., and Prudscova T.N., Phosphite-stabilizers (AO-118) impact on molecular weight characteristics of engineering plastics, *Bratislava International Conference on Polymers*, 1990.

[32] Gur'yanova V.V., Ershov O.V., Mednikova G.S., Sharyigina I.A., Kalugina E.V., Rheological properties of polysulfone mixtures with various molecular weight characteristics, *Plastmassy*, 1990, No. 6, pp. 82 – 84. (Rus)

[33] Blumenfeld A.B., Bolotina L.M., Narkon A.L., and Kalugina E.V., *USSR Patent No. 1,550,913*, The method of polysulfone extraction, 1988. (Rus)

[34] Narkon A.L., Kalugina E.V., Volkov M.A., and Astakhov P.A., New superconstructional plastics based on thermoresistant liquid-crystal polyesters, *Plastmassy*, 1996, No. 3, pp. 4 – 6. (Rus)

[35] Kalugina E.V., Novotorzeva T.N., Andreeva M.B., Tochin V.A., Gorilovsky M.I., Blumenfeld A.B., Narkon A.L., Gumargalieva K.Z., and Zaikov G.E., Liquid crystalline copolymers as modifiers for some thermoplastics. (in press)

[36] Kulagina E.V., Novotorzeva T.N., Andreeva M.B., Tochin V.A., Gorilovsky M.I., Gumargalieva K.Z., and Zaikov G.E., Thermal behavior and stabilization of aryl-aliphatic polyamide. (in press)

[37] *J. Polym. Sci., Polym. Plus. Ed.*, 1985, vol. **23**, No. 4, p. 521.

[38] *Liquid Crystalline Polymers*, Ed. N.A. Plate, 1988, Moscow, Khimia, 415 p. (Rus)

[39] Shcherbakova T.S., Chudina L.I., Myagkov M.V., and Semenova A.I., Thechnological features of polyamidoimides processing, *Plastmassy*, 1987, No. 7, pp. 25 – 27. (Rus)

[40] Andreeva M.B., Novotorzeva T.N., and Kalugina E.V., Degradation and stabilization of aliphatic-aromatic polyamides, *Proc. 11th Symposium "Modern Chemical Physics"*, September 18 – 29, 1999, MSU boarding house "Burevestnik", Tuapse. (Rus)

[41] Andreeva M.B., Saratovskaya N.V., and Kalugina E.V., Thermal transformations and stabilization of aliphatic-aromatic polyamide, *Proc. 2nd All-Russian Kargin's Symposium "Chemistry and Physics of Polymers at Early 2000ies* (with international participants), May 29 – 31, 2000. (Rus)

[42] Kalugina E.V., Novotorzeva T.N., Andreeva M.B., Blumenfeld A.B., Urman Ya.G., and Gur'yanova V.V., Thermal transformations of aliphatic-aromatic polyamide, *Plastmassy*, 2001, No. 1, pp. 20 - 22. (Rus)

[43] Buchachenko A.L., *Complexes of radicals and molecular oxygen with organic molecules*, 1970, Moscow, Nauka, 263 p. (Rus)

Chapter 11

KINETIC ASPECTS OF AGING OF POLYVINYLCHLORIDE BASED POLYMER MATERIAL

G. E. Zaikov,* K. Z. Gumargalieva,**
T. V. Pokholok* and Yu. V. Moiseev**
* Institute of Biochemical Physics, Russian Academy of Sciences
117334 Moscow, 4 Kosygin str.
** Institute of Chemical Physics, Russian Academy of Sciences
117334 Moscow, 4 Kosygin str.

1 INTRODUCTION

Polyvinylchloride-based polymers are widely used as insulating materials in wires and cables. Modern flying means possess tens of miles of wires and cables, and in some cases exploitation time of flying means is defined by time of wire application.

Displaying a number of undoubtful advantages, polyvinylchloride-based materials possess the drawback: they all are more or less subjected to aging.

Not enough stability to aging may cause serious consequences, so it requires the most occurate prognosis of the exploitation time of polyvinylchloride-based insulating material. Underestimation of the exploitation time leads to early repair of flying means. The consequence of overestimation of the explaitation time are unexpected repairs and accidents.

At present the list of GOSTs and OSTs [1, 2] devoted to the problem of prognosis is in use, simplified empyric methods of prognosis are set, which do not fit modern requirements.

Successful development of science on polymer aging allows us to work out a prognosis technique of exploitation times of polyvinylchloride-based insulating materials for flying means based on modern concepts of polymer material aging.

The present review is devoted to studies of the aging mechanism of PVC-placticate and working out methods of diagnostics and prognosis of the exploitation time of board electric network in flying means.

2 AGING OF POLYVINYLCHLORIDE

Aging of polyvinylchloride at processing and exploitation is the matter of multiple investigations. The most of investigations are devoted to the study of dehydrochlorination of pure polymer in an inert atmosphere, as a rule. By the way, PVC is most often used as plasticate in practice, i.e. mixed with plasticizers the amount of which may reach 70% of the total mass. All this makes the problem of PVC aging the extremely complex event and requires consideration of all processes proceeding under natural conditions of processing and exploitation.

In general case the following processes can proceed at PVC-plasticate aging:

- PVC dehydrochlorination;
- Thermoxidative degradation of PVC and plasticizers;
- Cross-linking;
- Diffusional desorption of a plasticizer from an article;

2.1 Dehydrochlorination of Polyvinylchloride

Dehydrochlorination reaction proceeds in PVC with a noticeable rate at high temperature range (> 150^{0}C) [3]:

$$CH_2 - CHCl - CH_2 - CHCl - CH_2 - CHCl \xrightarrow{-HCl}$$

$$\underset{(\beta\text{-chlorallyl group})}{CH_2 - CHCl - CH_2 - CH = CH - CHCl} \xrightarrow{-HCl} CH_2 - CH = CH - CH = CH - CHCl \rightarrow etc.$$

Scheme 2.1

Existence of hydrochloride, the reaction product, and oxygen in the surrounding accelerates dehydrochlorination [4, 5]. That is why it is worthwhile, first of all, to consider reaction of noncatalytic dehydrochlorination.

2.1.1 Noncatalytic Dehydrochlorination

Noncatalytic dehydrochlorination proceeds under conditions, when the rate of diffusional desorption of hydrochloride from a sample significantly exceeds the rate of the reaction itself. For example, this condition is fitted for PVC powder or thin films at carrying out experiments in vacuum or in inert gas.

The reaction is the order one by polymer. The correct data were obtained in solution [5]. In solid polymers physical structure changes during dehydrochlorination, what leads to the distortion of the reaction order [6].

The rate of noncatalytic dehydrochlorination $\left(W^0\right)$ is calculated by the equation:

$$W^0 = \frac{\alpha}{t}; \ \alpha = \frac{n_{HCl}}{n_{HCl}^{\infty}}. \tag{1}$$

Here n_{HCl} and n_{HCl}^{∞} are HCl amounts formed during time t and at complete polymer transformation, respectively. W^0 is constant until $\alpha = 0.3 \div 0.5$ in dependence on reaction conditions.

HCl extraction is accompanied by formation of double bonds (polyene structures) in macromolecules, what is displayed by occurrence of characteristic absorption bands in UV- and visible spectrum parts [7]. We also can observe correlation between α and optical density growth at low degrees of transformation [8].

Process of noncatalytic dehydrochlorination

The process of noncatalytic dehydrochlorination consists of three main stages:

- Initiation of active centers;
- Propagation of dehydrochlorination;
- Termination of active centers.

Initiation

Initiation of dehydrochlorination may proceed by the random law (β-chlorallyl group formation) or by active centers (defect structures) formed during synthesis and storage of PVC.

Starting from primary investigations on PVC degradation and till the present [9 – 15] it has been commonly accepted that the reason of low PVC stability comparing it with model low-molecular chloralkans is the existence of defect structures in macromolecules.

Double Bonds

Activating action of double bonds is confirmed by the decrease of dehydrochlorination rate at soft PVC chlorination [16]. Double bonds may be identified and quantitatively determined by bromination followed by determination of the number of breaks, ozonolysis [17 – 19] and NMR technique [20]. The following results have been obtained in this case (see Table 2.1) [21].

Table 2.1. The amount of double bonds of different types in industrial PVC samples.

Double bond types	Amount per 1000 monomer units
$-CH_2-CHCl-CH=CH_2$	$0.7 \div 2.2$
$-CH_2-CHCl-CHCl=CH_2$ or $-CH_2-CHCl-CH=CH-CH_2-$	$0 \div 2.8$
$-CH_2-CH=CCl-CH_2Cl$ or $-CH_2-CH=CCl-CHCl-$	$0.6 \div 2.7$

A satisfactory correlation between W^0 and amount of terminal double bonds has been obtained in some works, as well as with the amount of internal double bonds [22 – 24].

Activating effect of internal double bonds has been discovered by Brown *et al.* [25] in investigations of dehydrochlorination of specially synthesized vinylchloride copolymers with phenyl acethylene.

Groups Possessing Tertiary Carbon Atoms

Branched structures detach HCl much easier than normal chains do. Industrial samples contain 5 to 7 such structures per 1000 monomer units [12, 26]. The following structures were detected with the help of the NMR technique (see Table 2.2).

Table 2.2. The amount of branched structures of different types in industrial PVC samples [8].

Branched structure types	Amount per 1000 monomer units
Chloroactylic	4 ÷ 6
Chloroethylic	0.4 ÷ 2.4
Butylic	0.4 ÷ 1.6
Non-identified longer ones	0.2 ÷ 2.4

'Head-to-Head' Structure

$-CH_2-CHCl-CHCl-CH_2-$. The amount of structures of this type does not exceed 6 per 1000 monomer units [27]. A special polymer has been synthesizes elsewhere that possessed the 'head-to-head' structure. The polymer possessed lower temperature of dehydrochlorination initiation, but the maximum of the rate was reached at higher temperature.

Stereoregularity

Syndiotactic structures are less stable than isotactic ones [28 – 32]. The content of long polyenes increases at dehydrochlorination with the increase of syndiotacticity, however the number of active centers of dehydrochlorination decreases in this case.

Oxygen-Containing Groups

Compositions possessing double bonds can be oxidized, hence different oxygen-containing groups may exist in PVC. According to the data [33] PVC may possess ketoallyl groups, $-CH_2COCH=CH-$, in amounts 0.01 ÷ 0.25 per 1000 monomer units. Existence of these groups is shown by the presence of polymer chain breaks in alkaline solution. Linear dependence of W^0 on ketoallyl group concentration allowed the authors of those works to conclude about the deterministic role of ketoallyl groups on the initial stage of dehydrochlorination [34, 35]. However, the presence of ketoallyl groups in industrial polymers is said doubtful in some works [10, 11]. The presence of oxygen in the polymerization medium leads to the following reactions [36]:

$$\sim CH_2-CHCl\sim+O_2 \rightarrow CH_2-CHCl-O-O^{\bullet} \rightarrow CH_2O+HCl+CO+R^{\bullet}$$

Carbon oxide formed may co-polymerize with vinylchloride forming structures with ketone groups [37]:

$$\sim CH_2-CHCl-CO-CH_2-CHCl,$$

which reactivity in dehydrochlorination is not clear yet.

Polymeric vinylchloride peroxides may be formed at vinylchloride co-polymerization with oxygen [38]

$$\sim CH_2-\underset{\substack{|\\ Cl}}{C}H-O-O-CH_2-CH\sim,$$

or at the cross break of peroxide macroradicals [39]

$$\sim CH_2-\underset{\substack{|\\ Cl}}{C}H-O-O-\underset{\substack{|\\ Cl}}{C}H-CH\sim,$$

or at the chain transfer to monomer:

$$\sim CH=CCl-O-O-CH_2-CHCl\sim,$$

$$\sim CH_2-CHCl-O-OH.$$

As the concentration of peroxide groups in PVC macromolecules is low, their direct experimental determination is difficult. Iodometry method allowed to register 10^{-3} mole of peroxide per mole of VC for a polymer produced in oxygen, and 2.5×10^{-4} mole of peroxide per mole of VC for a polymer processed under 0.15 Pa vacuum [19, 20].

The proofs of existence of peroxide groups are:

- Absorption in 850 ÷ 900 and 1050 ÷ 1080 cm^{-1} ranges of the IR-spectra;
- Absorption in 350 ÷ 370 nm range of the UV spectrum;
- The decrease of PVC molecular mass at polymer processing by acids and alkali, that leads to heterolysis of peroxide bond in the presence of these substances:

$$-CH_2-\underset{\substack{|\\ Cl}}{C}H-O-O-\underset{\substack{|\\ Cl}}{C}H-CH-\rightarrow CH_2-C\overset{O}{\underset{Cl}{\lessgtr}}+OH-\underset{\substack{|\\ Cl}}{C}H-CH-.$$

Polyene Structures

According to data from [33] polyene structures are absent in PVC, but other data [17] show their concentration of 0.4 per 1000 monomer units.

Admixtures

Industrial PVC samples contain large amount of various admixtures: metals, solvent residues, emulgants, initiators. Many of them can accelerate PVC dehydrochlorination. This is testified by the increase of polymer stability (by 30%) at its complete purification. Among admixtures the influence of initiators on dehydrochlorination rate is most well studied. For example, W^0 grows with the increase of 1,1'-azobisisobutyronitrile concentration in PVC (from 10^{-2} to 1 mol%) [42, 43].

In this case great number of various defective structures is known, capable to initiate PVC dehydrochlorination. There are several types of defective structures existing in industrial polymers, apparently, it is difficult to find a correlation between W^0 and concentration of a single type of them. This point is confirmed by the data of the work [44].

Propagation of Dehydrochlorination and Active Center Termination

Propagation of dehydrochlorination proceeds with HCl detachment from β-chloroallyl group (allyl activation mechanism), or from chain fragments with the above mentioned defective structures. Double bonds formed, in their turn, activate HCl detachment from neighbour $\sim CH_2 - CHCl \sim$ groups (see the Scheme 1), that finally leads to the formation of polyene structures.

Comparing rate constants of dehydrochlorination in model PVC compositions with different amounts of double bonds we can see that introduction of each double bond increasing K^0_{eff} value by two decimal degrees, approximately, in both gas and liquid phases (see Table 2.3).

Table 2.3. Kinetic parameters of dehydrochlorination reaction of model PVC compositions in the gas phase [46].

Model composition	Temperature range, ^{0}C	$K_{177^0C}\times10^{11}$, s^{-1}	*E*, kJ/mol	lg*A*, s^{-1}
2-chlorbutane	347 ÷ 394	3.6	206	13
2,4-dichlorpentane	347 ÷ 397	3.3	206	13
2,4,6-trichlorheptane	346 ÷ 386	3.4	214	14
4-chlorpentene-2	296 ÷ 345	859	172	12
6-chlorheptadien-2,4	302 ÷ 347	45900	147	10
8-chlorhexadecane	236 ÷ 284	1	151	10.5
8-chlortridecene-6	165 ÷ 196	10^3	105	8.1
4-chlordodecene-2	156 ÷ 180	8×10^2	125	6.9
7-chlornonadyene-3,5	70 ÷ 96	4×10^5	81	8.0
6-chloroctadyene-2,4	87 ÷113	3.2×10^5	75	7.1

In such a manner the process of dehydrochlorination may be considered as an assembly of consecutive reactions, the first of which limits the rate of the whole process.

Polyene structures formed usually contain 4 to 10 double bonds [45]. It displays that termination of active center happens after formation of several double bonds instead of dehydrochlorination propagation along the whole molecule, the concentration of polyene structures being comparatively low in this case.

The mechanism of active center termination is not clear yet. It is probable that active center termination proceeds as a result of cyclic structure formation [47 – 49]. Steric obstacles, appeared at dechydrochlorination proceeding in solid polymer, must not be excluded from the list of probable reasons, as well.

Polyvinylchloride Stability

Stability of PVC in non-catalytic dehydrochlorination is most often estimated by W^0 value. The data obtained by different authors for the process of non-catalytic PVC dehydrochlorination under vacuum are successfully described by the Arrhenius equation (Fig. 2.1). The activation energy value is 134 ± 8 kJ/mol. The following equation [48 – 50] can be used for calculations in 90 ÷ 230°C temperature range:

$$W^0 = (2 \pm 1) \times 10^9 \exp\left(-\frac{134 \pm 8}{RT}\right) \ (s^{-1}). \qquad (2)$$

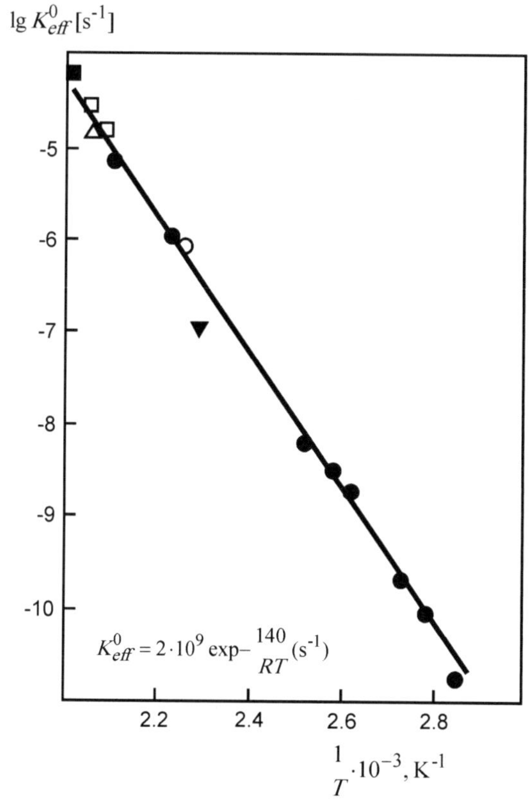

Fig. 2.1. The dependence of the rate constant logarithm of noncatalytic dehydroclorination on reverse temperature.

Two questions appear at estimation of PVC stability:

1. How does W^0, obtained for a polymer, differ from that of model compositions in the liquid state?
2. What difference exists for W^0 obtained for different industrial polymers.

Table 2.4 shows W^0 values for PVC and chloroalkans in the liquid phase at 177 and 200^{0}C.

Table 2.4. Values of W^0 for PVC and chloroalkans in the liquid phase.

Compound	$W^0_{177^0} \times 10^7, s^{-1}$	$W^0_{200^0} \times 10^6, s^{-1}$
PVC	$6 \div 9$	$2 \div 4$
8-chlorhexadecane	1	1
2,4-dichlorpentadecane	17	5
1,4,7-trichlorheptane	6.5	3
1,4,9-trichlornonane	6.0	4

It is seen that W^0 for PVC and chloralkans in the liquid phase dispose in the range of one decimal degree, the coincidence of rate constants of dehydrochlorination is improved with the increase of chlorine atomic amount in model compositions. However, we should deal with these data very carefully, because the influence of solvent on dehydrochlorination rate must be taken into account at the estimation of the reactivity of chloroalkans [51].

Table 2.5 shows W^0 values for different industrial polymers [52].

Table 2.5. W^0_{eff} values for different industrial polymers at 177^{0}C and 10^{-2} Pa.

Trade mark	State	$\bar{M}_v \times 10^{-3}$	$W^0_{eff} \times 10^7, s^{-1}$
S-70	Russia (former USSR)	138	8.2
S-68	Russia (former USSR)	104	7.2
S-55	Russia (former USSR)	70	10.3
Sircon 548 PM	Italy	143	7.1
Ravinyl PP 100/70P	Italy	136	7.0
TK-1300	Japan	163	6.8
B/III-8000	Japan	139	6.1
Nicavinyl-1300	Japan	132	7.1
Vestolit S 7064	Germany	130	6.9
Vestolit S 6051	Germany	90	8.8
Hostalit M-VP 3067	Germany	108	9.2
Lucovil-1150	France	133	9.0
Revicon R-45	Sweden	125	8.0
Solvic 223	Belgium	65	7.9

Industrial polymers from different countries display small differences in W^0_{eff}, and its average value is $(7.8 \pm 0.9)\times10^{-7}$ s^{-1}.

Thus, basing on the above mentioned facts as in the work [53] also, we may accept that stability of industrial PVC in dehydrochlorination is determined, on the whole, by structure of elementary chloroalkyl unit, and defective structures exist in these polymers in similar amounts or play no significant role.

As the process of dehydrochlorination is complex and includes several consecutive reactions, the attempts have been made in some investigations to determine kinetic parameters of these reactions.

Statistic Hcl Detachment from Chloralkan Chains

$$\frac{d[\mathrm{HCl}]}{dt} = K_c[\mathrm{ChA}], \tag{3}$$

here K_c is the rate constant of dehydrochlorination at HCl detachment from chloroaklane groups; $[\mathrm{ChA}]$ is the concentration of chloroalkane groups in polymers.

The rate of this reaction is mostly often determined by the accumulation of breaks in PVC macromolecules at ozonolysis [18, 19], and more rare by the NMR method by accumulation of internal double bonds [54]. Kinetic parameters of this reaction were calculated in the work [53]:

$$K_c = 10^{11\pm0.5}\exp\left(-\frac{160\pm40}{RT}\right)\mathrm{s}^{-1}. \tag{4}$$

Values of pre-exponential multiplicand and activation energy are close to the correspondent parameters of 8-chlorhexadecan (see Table 2.3).

HCL Detachment from Chloralkan Chains by Means of Activation by β-Chloroallyl or Ketoallyl Groups.

$$\frac{d[\mathrm{HCl}]}{dt} = K_I[\beta\text{-}\mathrm{ChA}]. \tag{5}$$

Here K_I is the rate constant of dehydrochlorination activated by β-chloroallyl; $[\beta\text{-}\mathrm{ChA}]$ is concentration of β-chloroallyl groups.

K_I values can be estimated from the data shown in [13]. K_I values at 150, 170 and 220^{0}C equal 7.5×10^{-5}, 2×10^{-4} and 1.2×10^{-4} s^{-1}, respectively.

$$K_I = (5\pm2)\times10^9\exp\left(-\frac{115\pm8}{RT}\right)\mathrm{s}^{-1}. \tag{6}$$

Kinetic parameters of dehydrochlorination initiated by ketoallyl groups were determined in the work [35]. Attempts were made to obtain general kinetic equations, which would take into account initiation of dehydrochlorination by different defective structures. Thus, for example, the concept about the deterministic role of ketoallyl groups was developed in [40] and corresponding equation was deduced. However this equation does not explain the experimental fact: the invariability of dehydrochlorination rate until transformation reaches its 0.3 ÷ 0.5 degree. Ketoallyl groups must be spent very fast because of their high reactivity. This is the reason of dehydrochlorination rate decrease.

The Mechanism of Noncatalytic Dehydrochlorination

There is no general opinion on the mechanism of noncatalytic dehydrochlorination yet [48, 49]. Ideas are spoken out for the benefit of the molecular-ionic as well as of the radical-chain mechanism during last twenty years.

The molecular-ionic mechaism provides for HCl detachment from a chloralkan group according to one of three different pathways [55, 56].

1. Monomolecular mechanism (E_1).

$$\sim CH_2 - CHCl \sim \leftrightarrow CH - \overset{+}{C}H \sim + Cl^- \text{ (slowly)}$$

$$\sim CH - \overset{+}{C}H + S \rightarrow CH = CH + SH^+ \text{ (rapidly)},$$

here S is a solvent.

Usually this mechanism is realized in ionizing solvents, and there are no bases to suppose any probability of such a mechanism proceeding in polymer matrix with dielectric constant ~ 4.

2. Bimolecular mechanism (E_2).

$$\sim CH_2 - CHCl \sim + B \leftrightarrow \sim \underset{\substack{| \\ H........B}}{C}H_2 - CHCl \sim \rightarrow \sim CH = CH + Cl^- + BH^+,$$

here B is a base.

According to this mechanism detachment proceeds in the presence of strong bases. Realization of such a mechanism is of a very low probability in industrial polymers.

3. Cyclic mechanism.

Cyclic detachment is a monomolecular single-stage process (contrary to the mechanism E_1), which consists of the fact that detached chlorine atom extracts β-proton with it when yields.

$$\sim \overset{\substack{H \\ |}}{C}H\overset{\substack{Cl \\ |}}{C}H \rightarrow \sim HC = CH \sim + HCl$$

This process is not ionic because it does not proceed through formation of ions in free state. But in some cases we may put forward an analog with ionic processes. For example, pyrolysis of alkylgaloids displays the same dependence of reactivity of the composition structure as in the mechanism E_1. This means that the transition state seems like ionic pair R^+Cl^- by its structure. For example, we can imagine that HCl detachment from chlorallyl group proceeds through formation of an allyl zwitterion:

$$\sim CH = CH - CHCl - CH_2 \sim\rightarrow \overset{\delta+}{C}HCH\underset{\underset{Cl}{|}}{\overset{\delta-}{C}}H - CH_2 \sim$$

Apparently, cyclic mechanism is the most suitable for non-catalytic dehydrochlorination of PVC.

The radical-chain mechanism includes formation of free radicals and their further dehydrochlorination by direct detachment of HCl or precedent formation of Cl atom. For the first time this mechanism was suggested in works [49].

The following arguments are usually put forward for the benefit of the molecular-ionic mechanism:

- Absence of formation of radical reaction products (H_2, Cl_2, etc.);
- HCl catalysis;
- Absence of influence of radicals' inhibitors;
- Decay of lowmolecular chloralkans according to the molecular-ionic mechanism.

To confirm the radical-chain mechanism of dehydrochlorination we can present data about the existence of the reaction of ^3H- and ^{14}C-labeled toluene introduction into a polymer chain and about correlation of the transformation degree of this reaction with the degree of PVC dehydrochlorination.

Recently the ion-radical mechanism of PVC dehydrochlorination was suggested [57], which includes C-Cl-bond ionization in a polymer segment with formation of an ion pair with cyclic chloronium ion, electron transfer and macroradical formation, what imitate dehydrochlorination. As to the authors' point of view the mechanism is confirmed by influence of thiols, dihydroanthracene and other proton donors on chain dehydrochlorination (macroradical formed is 'exhausted' by interaction with H).

2.1.2 Catalytic Dehydrochlorination

Hydrogen chloride plays a very important role in PVC aging. It was found that HCl catalyzes dehydrochlorination. PVC catalytic dehydrochlorination proceeds when the diffusion rate of formed hydrogen chloride is comparable with the rate of non-catalytic dehydrochlorination, or the reaction proceeds in hydrochloride atmosphere.

Hydrogen Chloride Solubility in PVC

Hydrogen chloride is soluble in PVC that is confirmed by kinetic data showing that the rate of catalytic reaction is proportional to polymer article volume [58]. However attempts to obtain quantitative data by gravimetric technique (McBain scales), undertaken in this work, failed.

Hydrogen chloride solubility in a polymer is determined in [59] by the monometric technique according to decrease of HCl pressure in a vessel possessing V_0 volume, which contains a definite mass of polymer (m). Solubility coefficient σ was determined by the following equation:

$$\sigma = \frac{P_0 V_0 - P_1 (V_0 - V)}{P_1}, \tag{7}$$

where P_0 and P_1 are initial and terminal (after polymer introduction) pressures of HCl, respectively; $(V_0 - V) = \Delta V$ is the system volume change.

The solubility obtained (σ_{20° = 6.6) is, apparently, overestimated, because in the case of PVC powders adsorption should preferably proceed. Specific surface of PVC powders is 0.3 to 1.6 m^2/g [60]. Nevertheless, the data obtained in the work are of a great interest, because they show that HCl solubility increases during degradation:

$$\sigma = \sigma_0 (1 + \beta \times \alpha), \tag{8}$$

here β is proportion coefficient equal 63 at 20°C.

Active centers of Catalytic Dehydrochlorination

In general case chloroalkan, β-chloroallyl and ketoallyl groups may be active centers in PVC catalytic dehydrochlorination [58, 61].

Chloroalkane groups

The role of chloroalkane groups as active centers of the catalytic dehydrochlorination is not clearly proved yet. Pocker [62] has found that HCl is a strong electrophilic catalyst in media with low dielectric constant, and it increases significantly the rate of tret-butylchloride dehydrochlorination in these media.

β-Chloroallyl groups are π-bases capable to attach HCl. The existence of β-chloroallyl groups-hydrogen chloride complexes was for the first time shown by Schlimper [63] *et al.* The role of these complexes in catalytic dehydrochlorination was stated in [3]. It was found that the rate of 4-chlor-2-hexane dehydrochlorination increases with the growth of solvent polarity (the condition favorable for the complex formation).

Table 2.6. Rate constants of 4-chlor-2-hexane dehydrochlorination in different solvents in presence of 0.8 $m \cdot l^{-1}$ of HCl at 60°C.

Solvent	Dielectric permeability	$K \times 10^4$ min^{-1}
Methanol	32.6	8
1,2-dichlorethan	10.3	3.5
Tetrahydrofuran	7.2	0
Dioxane	2.2	1.0
Heptane	2.0	0.8
Ethyl-2-hexyl	11.0	1.5

In accordance with the opinion displayed in the present review this complex is a complex with charge transfer.

Most strict proofs of β-chloroallyl group activity in catalytic dehydrochlorination were obtained by the ozonolysis with further counting the number of breaks in PVC after dehydrochlorination [64].

If β-chloroallyl groups are active centers of catalytic dehydrochlorination, then $\overline{M}_n$ of samples subjected to degradation in presence or in absence of HCl must be the same, because detachment of polyene fragments will not affect $\bar{M}_n$ value (the length of polyene fragments is much shorter than that of PVC fragments, which include nonreacted chloroalkane groups).

If chloroalkane groups are active centers of the reaction, $\bar{M}_n$ of PVC samples must change during degradation in presence of HCl, because formation of double bonds in PVC molecules will proceed by the random law.

The experiments performed at temperatures close to room temperature showed that the change of break amount in PVC samples with time subjected to degradation in presence or in absence of HCl are practically equal, i.e. under present conditions reactivity of β-chloroallyl groups is much higher than that of chloroalkane groups.

Groups with tertiary carbon atoms are n-bases and, consequently, may be active centers of dehydrochlorination. Structures possessing general formula $R_1R_2R_3CCl$ with tertiary C-Cl bond and substituents R_1, R_2 and R_3, which contain polyene fragments capable for conjugating with α-carbonic atom [65], must be the most basic ones.

Macrokinetics of PVC Catalytic Dehydrochlorination

General equation of catalytic dehydrochlorination is the following:

$$\frac{dc_{\text{HCl}}}{dt} = \frac{dc_{=}}{dt} = \sum_i K_{eff}^{k} c_{\text{HCl}} c_{(a.c.)} \cdot \tag{9}$$

Here c_{HCl} is HCl concentration in polymer; $c_{=}$ is concentration of double bonds formed; $c_{(a.c.)}$ is concentration of active centers in polymer.

At temperatures close to room temperature β-chloroallyl groups existed in the polymer are active centers because noncatalytic dehydrochlorination proceeds at a very low rate.

At low degrees of transformation and constant HCl pressure the reaction system fits the equation:

$$\alpha = K_{eff} t , \tag{10}$$

here $K_{eff} = K_{eff}^{k} c_{\text{HCl}}^{0} c_{a.c.}^{0} \cdot$

This equation was applied to describe experimental data on catalytic dehydrochlorination of thin PVC films (no diffusional limits) in HCl and in hydrochloric acid vapours at 60°C [58].

In [59, 66, 67] the reaction of catalytic dehydrochlorination was performed in a closed volume at high temperature. The reaction proceeding was checked by the increase of hydrogen chloride pressure in the system. Non-catalytic dehydrochlorination proceeds with noticeable rate at temperatures over 185°C, and concentration of active centers (β-chloroallyl groups) will increase during the reaction.

Thus, autocatalytic character of PVC dehydrochlorination is stipulated by two reasons: the increase of hydrogen chloride concentration in a polymer and growth of active center amount of catalytic dehydrochlorination.

General kinetic equation displays a complex image [58], however experimental data obtained in works [59, 66, 67] at low transformation degrees are successfully described by the following equation (see Fig. 2.2):

$$\lg \frac{\alpha}{1-\alpha} = \lg \alpha_0 + 0.434 K_{eff}^{k} c_{a.c.}^{0} t, \quad (11)$$

where α is a parameter connected to HCl pressure in the system.

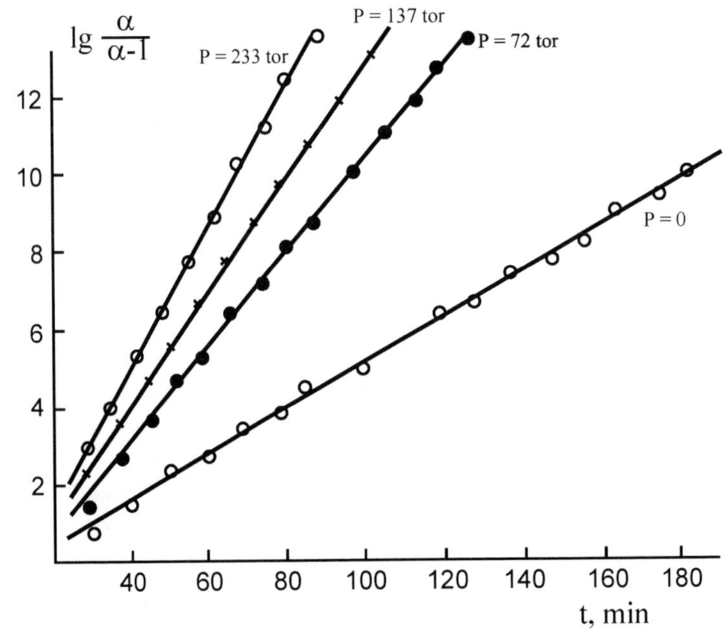

Fig. 2.2. Treatment of the experimental data of HCl pressure influence on the kinetics of dehydrochlorination at 200°C in accordance with Eq.(11).

Expressions $K_{eff}^{k} c_{a.c.}^{0}$ calculated for different initial HCl pressures for PVC powder dehydrochlorination at 210 °C are shown in Table 2.7.

Table 2.7. Values of $K_{eff}^{k} c_{a.c.}^{0}$ calculated by the equation (11) for PVC powder dehydrochlorination at 210 °C [58].

P_{HCl}^{0}, torr	0	72	137	233
$K_{eff}^{k} c_{a.c.}^{0} \times 10^4$, s^{-1}	1.0	2.0	2.3	2.8

The $K_{eff}^{k} c_{a.c.}^{0}$ change with the P_{HCl}^{0} increase may be due to the following reasons: nonobservance of the Henri law and the change of the active center-HCl complex concentration with the increase of HCl concentration in the polymer.

In accord with the data [40, 41, 48] the presence of HCl in the atmosphere catalyzes only detachment of hydrogen, making no effect on the length and distribution of polyene structures.

Kinetic regularities of PVC film dehydrochlorination at simultaneous proceeding of noncatalytic and catalytic reactions were considered in works [53].

The change of HCl concentration in a polymer in diffusional-kinetic range during dehydrochlorination can be described by the following equation:

$$\frac{\partial c_{HCl}}{\partial t} = D\frac{\partial^2 c_{HCl}}{\partial x^2} + K_{eff}^{0} + K_{eff}^{k} c_{a.c.}^{0} c_{HCl}, \qquad (12)$$

where D is the coefficient of HCl diffusion in a polymer.

The equation possesses an analytical solution if we accept that the concentration of active centers at catalytic dehydrochlorination is constant ($c_{a.c.} \approx c_{a.c.}^{0}$). The equation (12) obtains the following solution:

$$c_{HCl} = -\frac{2K_{eff}^{0}}{\pi}\sum_{n=1}^{\infty}\frac{1-\cos n\pi}{n\left(K_{eff}^{k} c_{a.c.}^{0} - Dn\pi^2/e^2\right)} \times$$

$$\times\left[1-\exp t\left(K_{eff}^{0} c_{a.c.}^{0} - Dn\pi^2/e^2\right)\right]\sin\frac{n\pi x}{l}, \qquad (13)$$

here l is the film thickness.

The amount of HCl diffused through specific surface during dt equals

$$dm = -D\left[\frac{dc_{\text{HCl}}}{dx}\right]_{x=l} dt = \frac{2DK_{eff}^{0}}{l} dt \sum_{n=1}^{\infty} \frac{(1-\cos\pi n)\cos n\pi}{\left(K_{eff}^{k} c_{a.c.} - Dn^2\pi^2 / l^2\right)} \times$$

$$\times\left[1 - \exp t\left(K_{eff}^{k} c_{a.c.} - Dn^2\pi^2 / l^2\right)\right]. \tag{14}$$

Final solution for the transformation degree α is the following:

$$\alpha = \frac{2DK_{eff}^{0}}{l^2 m_{\infty}} dt \sum_{n=1}^{\infty} \frac{1-\cos n\pi}{\left(K_{eff}^{k} c_{a.c.}^{0} - Dn^2\pi^2 / l^2\right)^2} \times$$

$$\left[\exp t\left(K_{eff}^{k} c_{a.c.}^{0} - Dn^2\pi^2 / l^2\right) - t\left(K_{eff}^{k} c_{a.c.}^{0} - Dn^2\pi^2 / l^2\right) - 1\right], \tag{15}$$

where m_{∞} is theoretical amount of HCl that may be extracted from a specific polymer volume.

Analysis of the equation (15) shows that there are two limit cases:

1. $K_{eff}^{k} c_{a.c.}^{0} >> D\pi^2 l^{-2}$, dehydrochlorination proceeds in autocatalytic regime, and sharp increase of the process rate is observed;
2. $K_{eff}^{k} c_{a.c.}^{0} < D\pi^2 l^{-2}$, acceleration of dehydrochlorination is not observed, and the maximum rate is reached at the time when exponential term of the equation becomes negligibly small.

Fig. 2.3 shows typical change of the transformation degree α with the time for dehydrochlorination of PVC samples of different thickness. As it is seen from the data shown in Fig. 2.3 the process rate increases with the growth of sample thickness. Pudov [59] was the first who introduced the idea of critical sample thickness, i.e. the value l at which degradation transfers from stationary regime to autoaccelerated one. Critical value l_{cr} is determined from the following condition:

$$\frac{\pi^2 D}{l_{cr}} = K_{eff}^{k} c_{a.c.}^{0}. \tag{16}$$

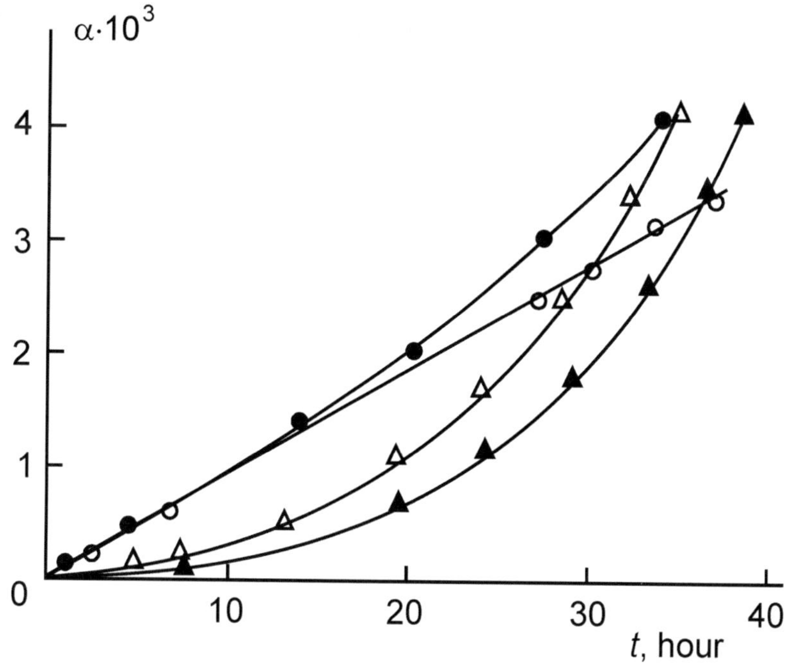

Fig. 2.3. The dependence of the transformation degree of hydrogen chloride on the time of dehydrochlorination of PVC disks of different thicknesses: (O) 0.05 cm; (●) 0.1 cm; (Δ) 0.2 cm; (▲) 0.4 cm at 140°C.

Table 2.8 shows values of $K_{eff}^{k} c_{a.c.}^{0}$ and D obtained by different authors.

Table 2.8. Values of $K_{eff}^{k} c_{a.c.}^{0}$ and *D* obtained for different temperatures [58].

t, °C	$K_{eff}^{k} c_{a.c.}^{0} \times 10^{4}$, s^{-1}	*D*×10, cm^{2}s^{-1}
120	0.075 ± 0.005	3.5 ± 0.5
130	0.190 ± 0.020	5.5 ± 0.5
140	0.440 ± 0.020	7.5 ± 0.5
180	1.000	
185	1.100	
200	1.700	33
210	1.700	

Activation energy of catalytic dehydrochlorination is 117 ± 8 kJ/mol.

Mechanism of Catalytic Dehydrochlorination

At present time there are several approaches to the mechanism of PVC catalytic dehydrochlorination. Generally acknowledged is the one of participation of polyene sequences in catalytic dehydrochlorination of PVC, and in this case it is accepted [61] that π-complex is reactive particle:

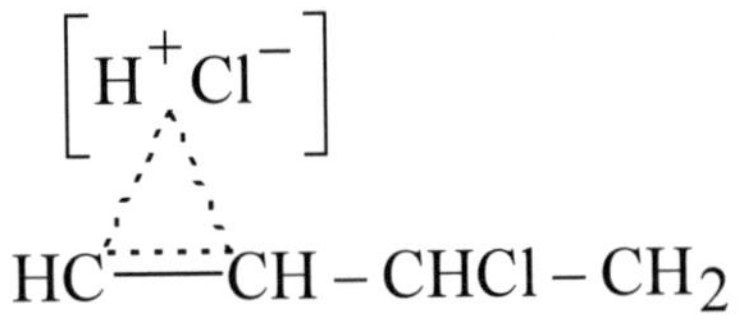

Such π-complex forms no carbonium ion, but promotes HCl detachment from neighbour chloroalkane group.

The mechanism of chain dehydrochlorination through polyene-cation system was suggested [69, 70] basing on rapid isotopic exchange between PVC and ^{3}HCl disposed in the atmosphere:

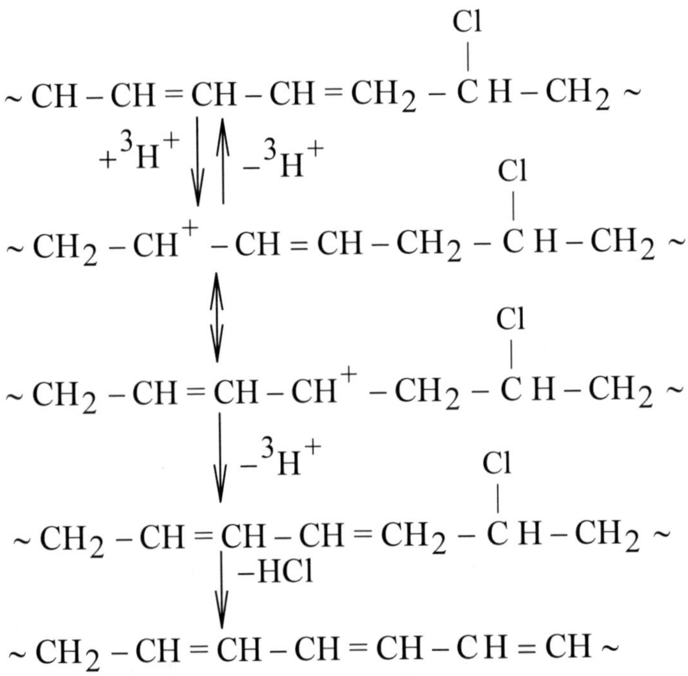

2.1.3 Plasticizer Effect on Dehydrochlorination

Plasticizer in PVC can exist in several forms:

- The real solvent;
- Adsorptional layer on the surface of overmolecular structures;
- Quasi-free phase in gaps between over-molecular structures.

Different forms of plasticizer can be identified, for example, by DTA [71] or NMR [72] techniques. It should be expected that plasticizer existed in different forms will differently affect the rate of dehydrochlorination; in its turn, chemical nature and concentration predict a ratio between plasticizer forms.

Fig. 2.4 shows typical picture of dehydrochlorination K_{eff}^{0} change from infinitely diluted PVC solution in plasticizer to pure polymer [73]. Proportion between K_{eff}^{0} and polymer concentration is usual, as a rule, in the range of diluted solutions (< 100 g/l). The range of concentrated polymer solutions does also show the increase of W^{0} with PVC concentration growth. In the range of intermediate concentrations some cases of W^{0} independence on polymer concentration changes were observed [74, 75].

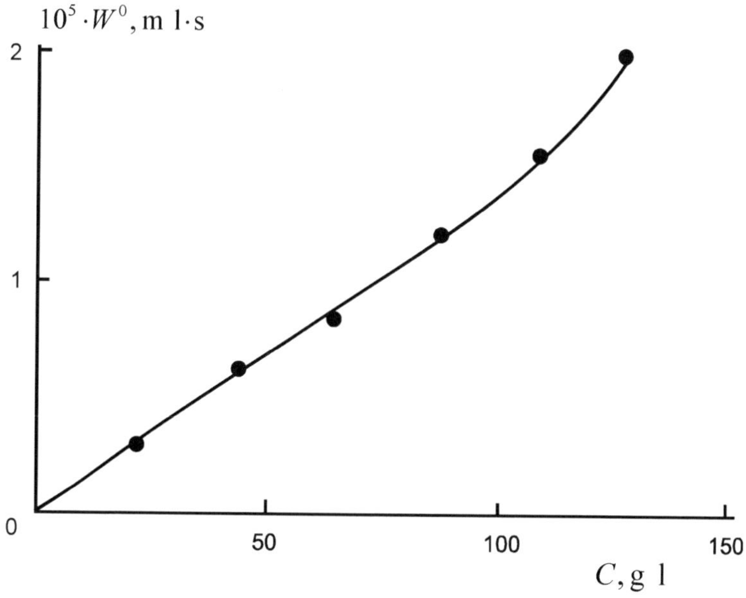

Fig. 2.4. The dependence of the rate of PVC dehydrochlorination in dioctylphthalate solution on the polymer concentration at 212°C.

Performing investigations in diluted solutions in this way, we can obtain information on plasticizer influence on PVC dehydrochlorination on molecular level. The data shows that nature of plasticizer influences the rate of dehydrochlorination in a significant manner, the value of W^{0} decreases, as a rule, with solvent polarity [74, 75].

Fig. 2.5 shows the dependence of the relative rate logarithm on reverse dielectric permeability for PVC dehydrochlorination in eight solvents polarity [74, 75]. Two of solvents (benzophenon and decalin) do not fit the present dependence. It is interesting that solvent polarity growth causes no increase of the activation energy, as it should appear in accordance with the compensation effect, but it does decrease (from 134 kJ/mol for PVC to 95 kJ/mol for benzyl alcohol). This induces the supposition that medium polarity is not the only factor that influences the dehydrochlorination rate. Probably, an important role is played by solvent acidity, what is observed, indeed, in practice [76, 77]. These are the bases from which the medium influence on attachement of hydrogen chloride to olefins is considered [3].

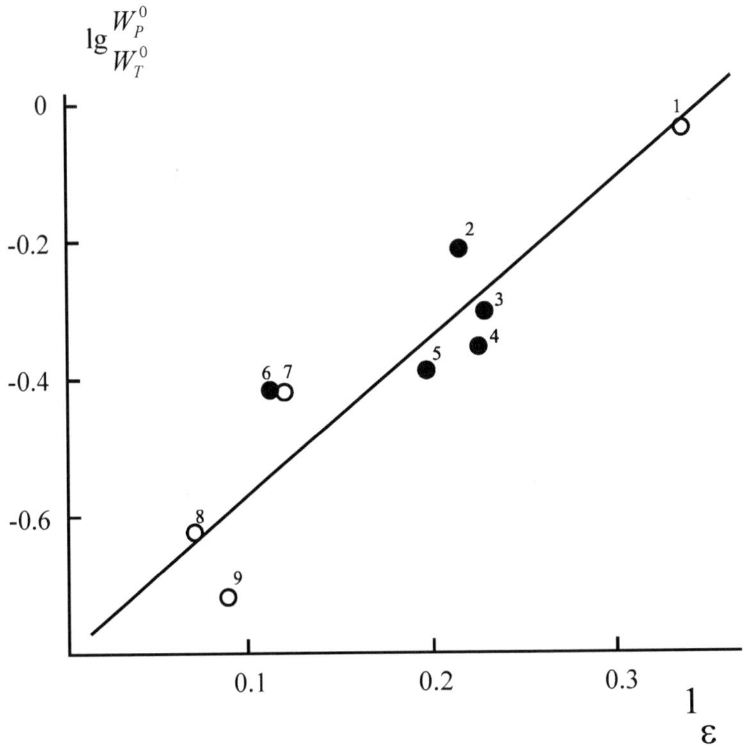

Fig. 2.5. The dependence of lg (W^o/W^o) on the reverse dielectric permeability for PVC dehydrochlorination in different solvents: 1 – PVC; 2 – tritholylphthalate; 3 – ethylbenzene; 4 – trioctylphosphate; 5 – dioctylphthalate; 6 – dichlornaphthalene; 7 – cyclohexan; 8 – benzylic alcohol; 9 – dichlorbenzene. The PVC concentration changes from 0.4 to 2.1 mass.%. W^o and W^o are dehydrochlorination rates in solution and in the solid phase, respectively.

Apparently, plateau in the dependence W^0 - C_{PVC} in the range of concentrated solutions is connected to the formation of a quasi-free plasticizer phase, and that plasticizer molecules disposed in emptinesses of over-molecular structures will not significantly influence the dehydrochlorination rate.

2.2 Thermooxidative Degradation of Polyvinylchloride

The following processes proceed during thermooxidative PVC degradation:

- Oxygen absorption;
- Hydrogen chloride extraction;
- Extraction of various volatile products;

- Decrease of mass and molecular mass of the polymer.

The correspondent kinetic curves are shown in Fig. 2.6 [81].

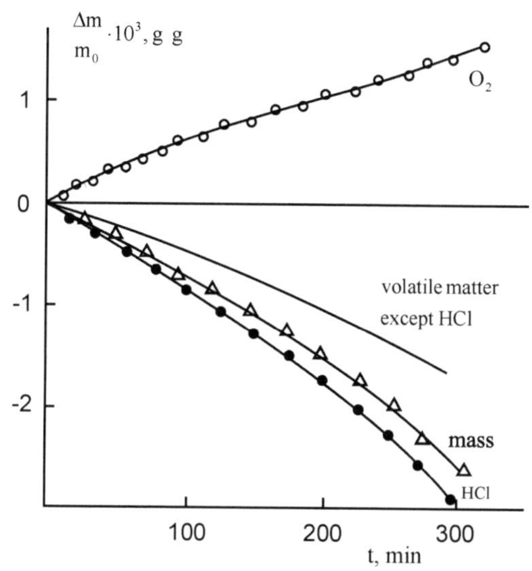

Fig. 2.6. Relative change of polymer mass at oxygen absorption, extraction of HCl and other volatile products during thermooxidative degradation of PVC at 180°C and 1 atm O_2. (From Ref.81.)

2.2.1 Oxidation of Polyenes

Polyene fragments of PVC are most easily oxidized. Kinetics of polyene oxidation has been studied in detail by N.M.Emanuel *et al.* [82, 83]. Direct attachment of O_2 to polyene in the basic triplet state is a spin-prohibited event. Oxidation proceeds through the stage of peroxyradical interaction with polyene structures with further formation of hydro- and cyclic additives:

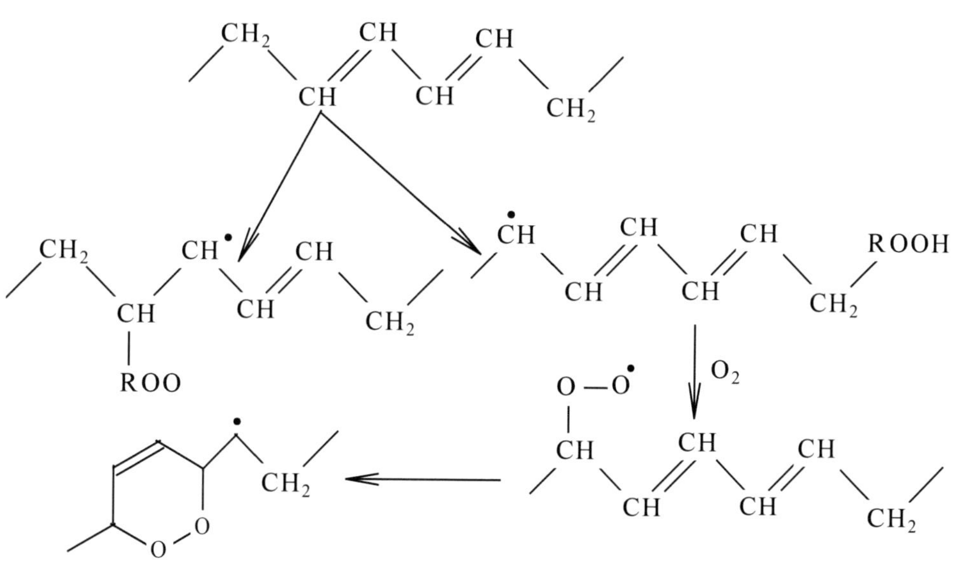

Decomposition of polyene radical proceeds with formation of a carbonyl compound and an oxyradical, the latter transforms into an allyl radical:

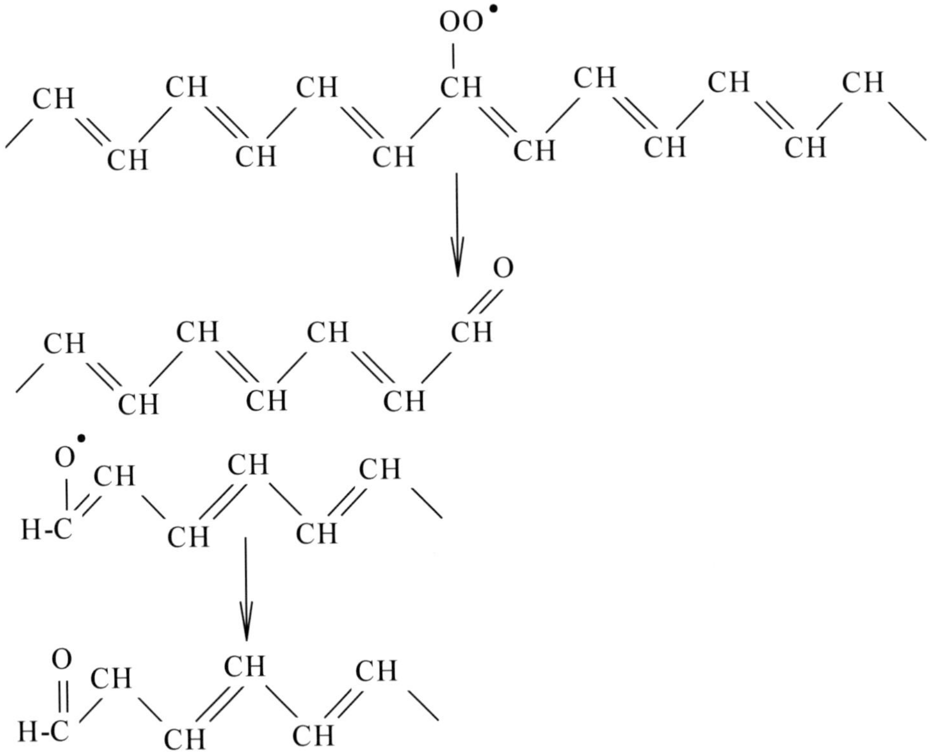

From the presented scheme it could be seen that the products are formed in reactions of isomerization and polyene peroxyradical decomposition, which contain smaller amount of double bonds. This has been confirmed by multiple investigations [49, 84, 85].

The rate of oxygen absorption is determined from the equation:

$$W_{O_2} = a[A]\sqrt{W_i}\frac{K_2}{\sqrt{K_6}} + \sqrt{W_i}\frac{K_x}{\sqrt{K_6}}, \tag{17}$$

here $[A]$ is concentration of polyenes; W_i is the initiation rate; K_2 and K_6 are rate constants of propagation and termination of chains, respectively. $\frac{K_x}{\sqrt{K_6}}$ characterizes relative reactivity of polyene peroxyradical in reactions of decomposition and isomerization.

PVC chain decomposition may, probably, proceed by the following scheme:

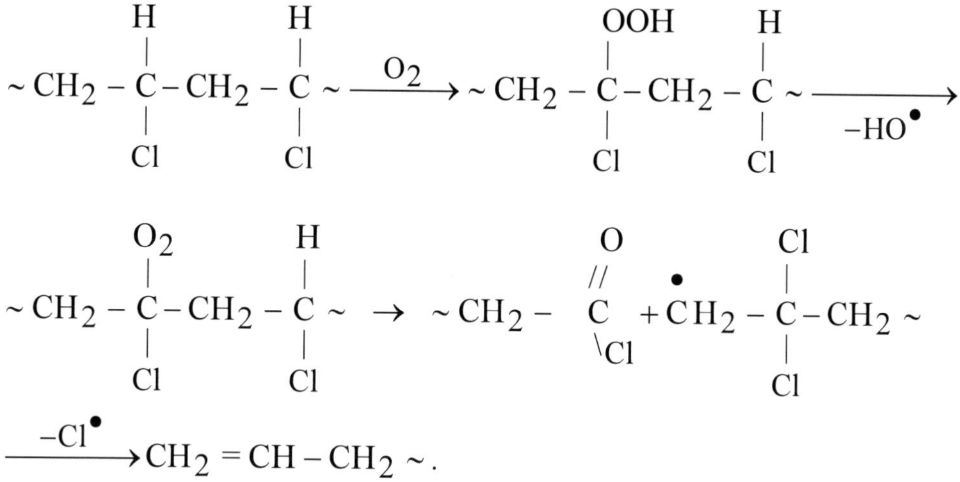

2.2.2 Oxygen Effect on Dehydrochlorination

Dehydrochlorination is significantly accelerated in presence of oxygen [48]. This is connected with formation of oxygen-containing groups, which initiate dehydrochlorination:

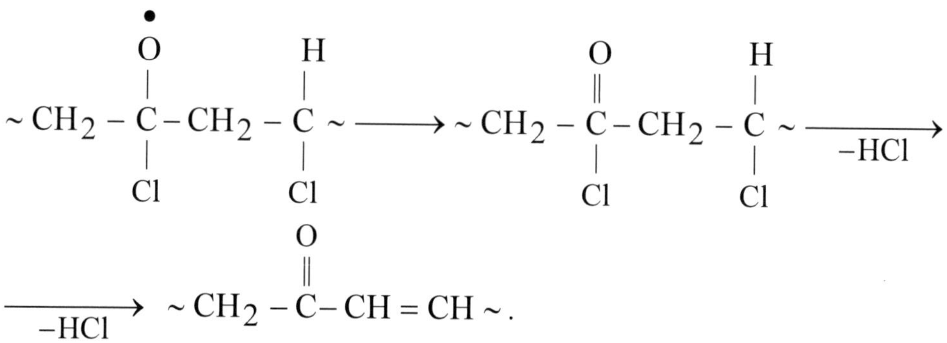

The change of the transformation degree α with time in argon/oxygen mixture. Kinetic curve of HCl yield is described by the formal equation [81, 86]:

$$\frac{d\alpha}{dt} = \left(K_s + K_\alpha\right)\left(1 - \frac{\alpha}{\alpha_\infty}\right), \tag{18}$$

where K_s is the rate constant of the initial part of the kinetic curve; K_α and α_∞ are formal constants.

Values of K_s at different temperatures are determined from the following equation:

$$K_s = 9 \times 10^8 \exp\left(-\frac{126}{RT}\right)\ \text{s}^{-1}. \tag{19}$$

2.2.3 *Effects of Plasticizers on Thermoxidative Degradation*

Influence of plasticizers accelerate thermoxidative degradation of PVC [87].

Fig. 2.7 shows kinetic curves of PVC dehydrochlorination, which contains different amounts of plasticizer.

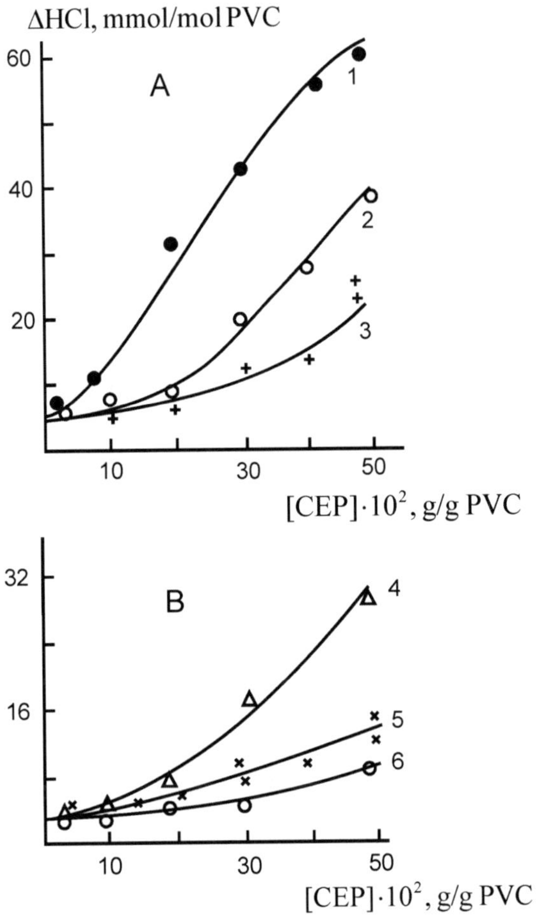

Fig. 2.7. The dependence of the hydrogen chloride amount yielded during 60 min on the concentration of a complex-ester plasticizer (CEP) at the thermodegradation of PVC (175°C) in O_2 (A) and in air (B): curves 1 and 4 – dioctylsebacynate (DOS); curves 2 and 5 – diisooctyladipinate (DOA); curves 3 and 6 – dioctylphthalate (DOP).

Autocatalytic character of PVC dehydrochlorination is bound to oxidation of plasticizers themselves and the influence of oxidation products on dehydrochlorination.

Acids and olefins are formed during thermodegradation of complex-ester plasticizers:

$$R\overset{\overset{O}{\|}}{C}OCH_2CH_2R^{\bullet} \rightarrow R\overset{\overset{O}{\|}}{C}O + R^{\bullet}CH_2\dot{C}H_2$$

$$R\overset{\overset{O}{\|}}{C}O^{\bullet} + RCOCH_2CH_2R^{\bullet} \rightarrow R\overset{\overset{O}{\|}}{C}OH + RCOCH_2\overset{\bullet}{C}HR^1$$

$$R\overset{\overset{O}{\|}}{C}OCH_2\overset{\bullet}{C}HR^1 \rightarrow R\overset{\overset{O}{\|}}{C}O^{\bullet} + R^{\bullet}CH=CH_2$$

$$R\overset{\overset{O}{\|}}{C}OCH_2\overset{\bullet}{C}HR^{\bullet} + RC\overset{\bullet}{H_2}\overset{\bullet}{C}H_2 \rightarrow R\overset{\overset{O}{\|}}{C}OCH_2CH_2R^{\bullet} + R^{\bullet}CH=CH_2.$$

Beside acids, compositions with peroxide, carbonyl and hydroxyl groups are formed in the presence of oxygen. These groups are those which initiate PVC dehydrochlorination. The mechanism of this process is not, however, quantitatively described.

It is known that unsaturated ketones activate dehydrochlorination according to the following scheme:

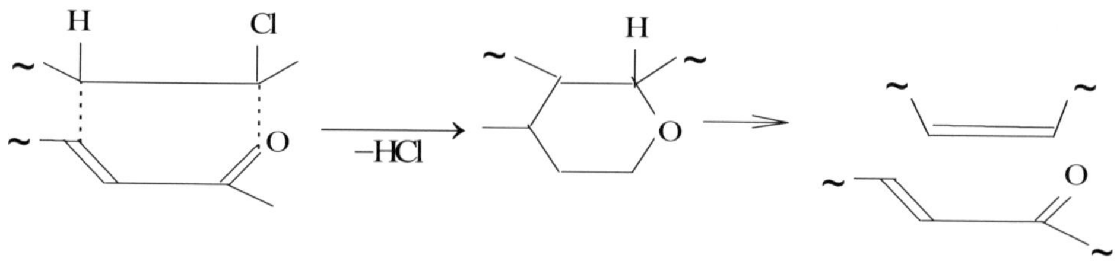

2.3 Cross-linking of Polyvinylchloride

Cross-linking of PVC macromolecules proceeds under same conditions as dehydrochlorination. Formation of cross-links and, at least, three-dimensional net structure leads to the increase of the molecular mass and the decrease of solubility (gel-fraction formation). Occurrence of gel-fraction happens after the induction period, the value of which depends on temperature (see Fig. 2.8) [88, 89].

The process of cross-sectional bond (*C*) formation may be presented as follows:

$$\frac{dC}{dt} = K_{cr.l.} c_{a.c.}^{n}, \tag{20}$$

where $K_{cr.l.}$ is the rate constant of the cross-linking; $c_{a.c.}$ is the concentration of active centers in macromoleculaes; *n* is the reaction order.

Concentration of cross-links at the moment of gel-formation (τ_2) is:

$$C_{cr.} = K_{cr.l.} c_{a.c.}^{n} \tau_2. \tag{21}$$

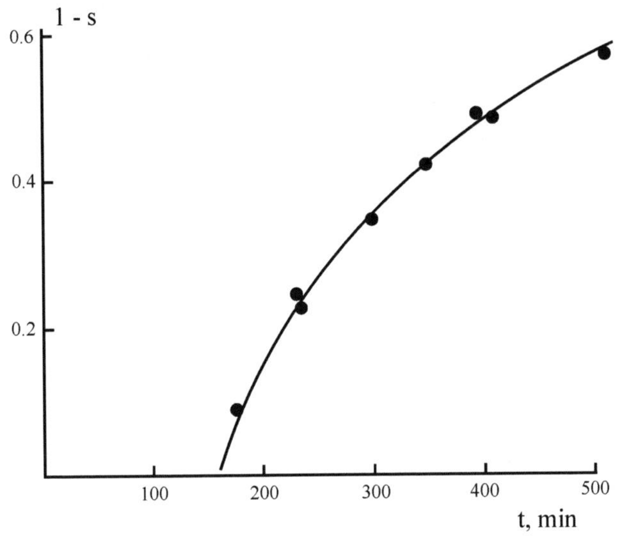

Fig. 2.8. Kinetic curve of gel formation during PVC thermooxidative degradation at 180°C in O_2. 1-s is the mass part of the gel.

Critical concentration of cross-links required for gel occurrence is determined by the known Flory expression:

$$C_{cr.} = \frac{1}{\bar{P}_{W^0}}. \tag{22}$$

Taking into account that mostof polymers possess $\bar{P_W} / \bar{P_n} \approx 2$, we obtain the following expression for $K_{cr.l.}$:

$$K_{cr.l.} = \frac{1}{2c^n_{a.c.}\tau_2 \bar{P}^0_n}. \tag{23}$$

PVC cross-linking is the order two reaction ($n = 2$) in both concentrated solutions in inert atmosphere and in solid polymer under vacuum. Temperature dependence of solid polymer cross-linking is described by the following equation:

$$K_{cr.l.} = (1 \pm 0.2) \times 10^7 \exp\left(-\frac{125}{RT}\right) \text{ s}^{-1}. \tag{24}$$

The activation energy of cross-linking in solution is ~ 100 kJ/mol [5].

Some authors [48, 90] consider cross-linking as the process proceeding with no dependence on dehydrochlorination, some others think that cross-linking is a secondary reaction with active centers presented by preliminarily formed polyene structures, and the reaction proceeds acording to the Dyllce-Alder dyene synthesis.

In presence of oxygen cross-linking proceeds simultaneously with degradation of macromolecules [85, 89]. The following equation is used to describe these reactions [91]:

$$S + \sqrt{S} = \frac{W_d}{W_{cr.l.}} + \frac{1}{W_{cr.l.}\, \bar{P}_n^0} \times \frac{1}{t}, \tag{25}$$

where S is soluble polymer fraction; W_d and $W_{cr.l.}$ are reaction rates of degradation and cross-linking of macromolecules, respectively.

Fig. 2.9 shows experimental data of PVC transformation in Ar and O_2, treated in accordance with equation (25). It is clearly seen that in inert atmosphere the reaction of macromolecule degradation does not take place. The value of W_d in O_2 atmosphere at 180°C is found in $(2 \div 20)\times10^{-8}$ s^{-1} range for various industrial PVC samples, and $W_{cr.l.}$ is found in $(5 \div 35)\times10^{-8}$ s^{-1} range [85].

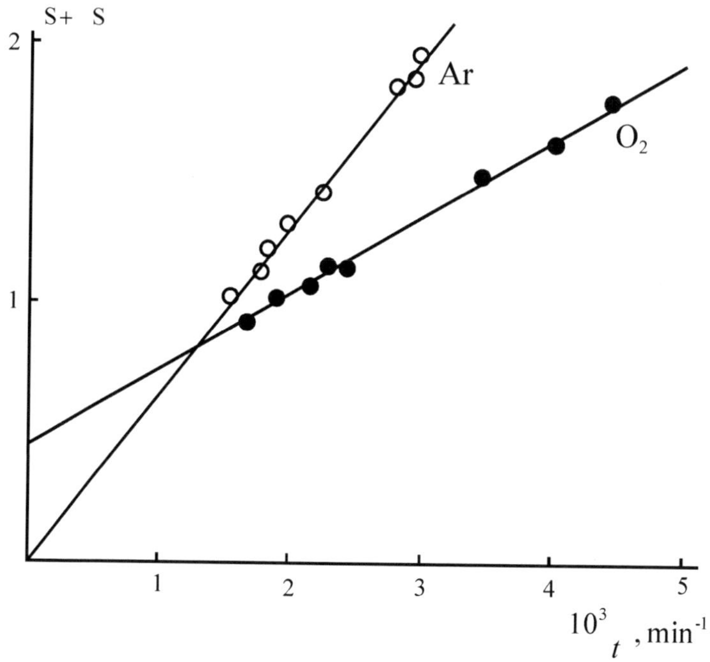

Fig. 2.9. Experimental data of gel formation at PVC degradation in Ar and O_2 at 180°C, treated in accordance with Eq. (25).

2.4 Formation of Volatile Products at Thermodegradation

Volatile product yielding at thermodegradation of PVC starts up at temperatures > 210°C. It is proved now that the following products are formed: benzene, naphthalene, styrene, toluene, methylnaphthalene [49, 92 – 94]. The first four compounds are formed in intramolecular reactions, whereas toluene and methylnaphthalene are products of intermolecular reactions with one or two H-atoms transfer.

Benzene formation is most well-studied reaction. Fig 2.10 shows the kinetic curve of this reaction at 180°C in air. The rate of benzene formation is 2.6×10^{-8} s^{-1}. There are two mechanisms suggested for this reaction [12, 95]:

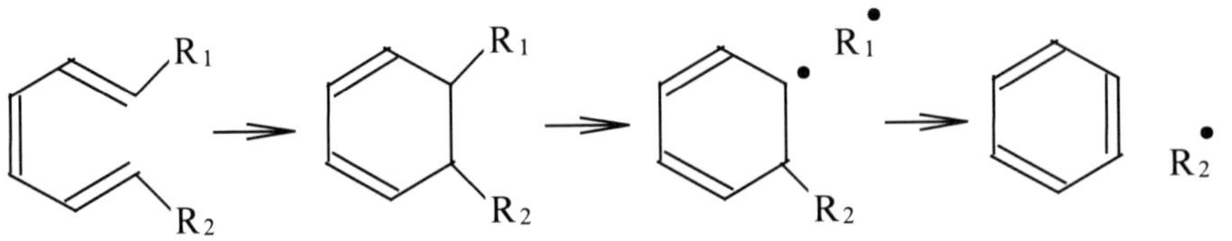

Then recombination of R_1 and R_2 radicals happens:

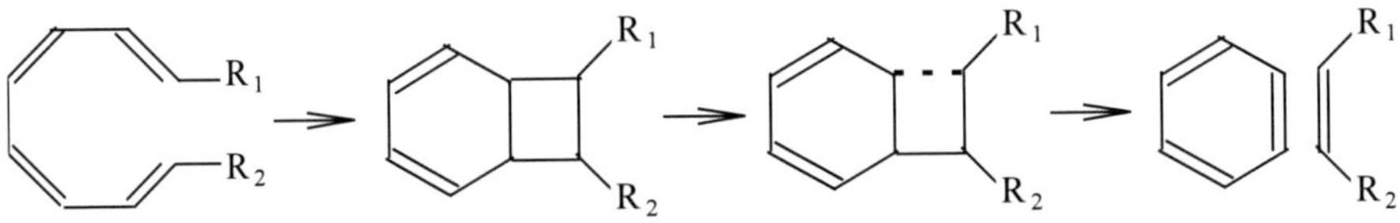

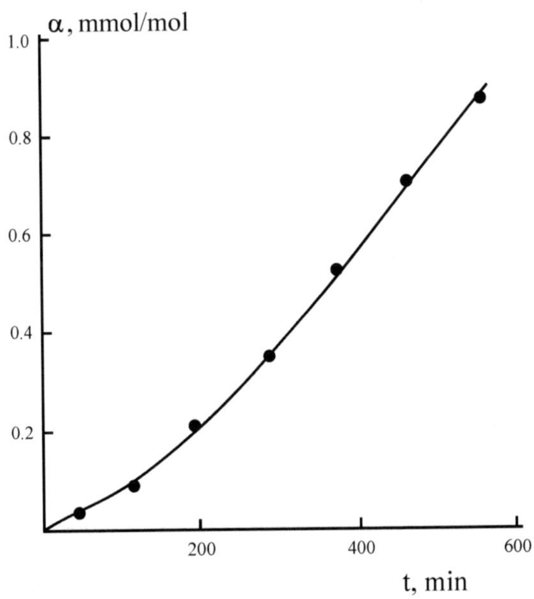

Fig. 2.10. Kinetic curve of the benzene formation at PVC degradation in Ar at 180°C.

2.5 Diffusional Desorption of Plasticizers from Articles

Processes of plasticizer diffusional desorption from PVC-plasticates are not well studied yet. Apparently, this is stipulated by the fact that under real conditions these processes proceed with low rates, dependent on geometrical sizes and construction of an article as well as on its prehistory. It is natural that these circumstances complicate comparison of data shown in different works. It is advisable to consider the existed data according to three following aspects:

- Dependence of plasticizer diffusional desorption on concentration and type of plasticizer;
- Time dependence of plasticizer diffusional desorption;
- The change of PVC-plasticizer properties at aging.

2.5.1 The Dependence of Plasticizer Diffusional Desorption on Concentration and Type of Plasticizer

Diffusion coefficients of different plasticizers were determined by gravimetric technique in vacuum [96]. Further on the data were concretized with the help of the refractional technique, because a correlation ratio exists between plasticizer concentration and refraction index [96]. Table 2.9 shows values of diffusion coefficients of plasticizers at different concentrations and temperatures. Table 2.10 shows values of the activation energy and constant D_0. It is clearly seen that diffusion coefficient as well as diffusion activation energy depend on plasticizer concentration.

Table 2.9. Diffusion coefficients of plasticizers in PVC (m^2/s).

Temperature, °K	Average plasticizer concentration, mass %	Tri-octyl phthalate	Di-octyl phthalate	Di-butyl phthalate	C_7-C_9 The ester of adipic acid and alcohols
298	80	3.8×10^{-23}	6.4×10^{-13}	1.9×10^{-12}	—
	60	3.0×10^{-13}	6.2×10^{-14}	4.1×10^{-13}	—
	40	5.0×10^{-14}	—	2.9×10^{-14}	—
313	80	6.5×10^{-12}	1.7×10^{-12}	3.1×10^{-12}	—
	60	1.6×10^{-12}	2.8×10^{-13}	1.3×10^{-12}	—
	40	1.8×10^{-13}	2.4×10^{-14}	2.1×10^{-13}	—
	10	—	—	—	2.1×10^{-15}
333	80	1.3×10^{-11}	3.2×10^{-12}	8.1×10^{-12}	—
	60	4.1×10^{-12}	1.1×10^{-13}	3.8×10^{-12}	—
	40	9.1×10^{-13}	1.1×10^{-13}	1.7×10^{-12}	—
	10	—	2.4×10^{-14}	2.4×10^{-15}	3.0×10^{-14}
353	80	1.8×10^{-11}	6.7×10^{-12}	1.6×10^{-11}	—
	60	6.4×10^{-12}	3.3×10^{-12}	1.3×10^{-11}	—
	40	1.3×10^{-12}	6.1×10^{-13}	3.3×10^{-12}	—
	10	8.7×10^{-15}	6.1×10^{-15}	4.0×10^{-14}	1.3×10^{-13}
373	80	3.0×10^{-11}	1.4×10^{-11}	3.2×10^{-11}	—
	60	1.2×10^{-11}	9.0×10^{-12}	2.3×10^{-11}	—
	40	5.6×10^{-12}	2.1×10^{-12}	1.2×10^{-11}	—
	10	5.3×10^{-14}	5.6×10^{-14}	2.6×10^{-13}	7.3×10^{-13}

Table 2.10. Diffusion activation energy and diffusion coefficients of plasticizers in PVC.

Plasticizer	Average plasticizer concentration, mass %	E, J/mol	D_0, m²/s
Trioctylphthalate	80	24.28	7.3×10^{-8}
	60	37.26	2.0×10^{-6}
	40	50.52	4.3×10^{-4}
	10	100.50	5.6×10^{-4}
Dioctylphthalate	80	36.43	1.7×10^{-6}
	60	56.52	6.8×10^{-4}
	40	74.53	5.9×10^{-2}
	10	126	2.3×10^{-4}
Ester of adipic acid and alcohols C_7-C_9	80	36.01	3.5×10^{-6}
	60	47.73	1.1×10^{-4}
	40	65.31	1.7×10^{-2}
	10	85.41	6.2×10^{-1}

Attempts to set quantitative connection between the value of evaporated plasticizer and its vapour pressure gave no satisfactory results, that in the case of high plasticizer yields is stipulated by deterministic value of the diffusion rate of plasticizer in the polymer. At the same time the comparison of different plasticizers shows correlation between pressure of their vapours and their evacuation from PVC-plasticate (see Table 2.11).

Table 2.11

Plasticizer	Vapour pressure, mm.Hg$\times10^6$	Mass loss after 100 hour exposure at 60°C, %
Tricrezylphosphate	8.2	0.6
Dibutylphthalate	1720	26.3
Dioctylphthalate	15.8	1.4
Dibutylsebacynate	326	15.0
Dioctylsebacynate	2.8	0.5

It was stated that the dependence of mass losses (%) on reverse film thickness for some time periods is expressed by straight lines.

The character and degree of different plasticizer migration from PVC-plasticates into insulating rubber vulcanize at room temperature during 26 months were investigated in the work [97]. The results obtained allowed a conclusion that better volatile plasticizers migrate much easier (see Table 2.12).

Table 2.12. Migration of plasticizers from PVC-plasticates at 20°C

Plasticizer	Migration into rubber vulcanisate, mg/dm^2	Volatility, mg/dm^2
Di-butylphthalate	390	228
Mesamol	255	169
DOP + DBP (3:1)	215	91
DOP	180	72
TCP + DBP (1:1)	123	72
TCP	0	0

It is seen that migration of plasticizer mixture is not an additive value, and less it is the value calculated according to the additivity rule.

Quantitative connection between effectiveness and mobility of plasticizer was set in the work [96]. Plasticizers possessing higher effectiveness (significantly decreasing vitrification temperature) are characterized by higher mobility in PVC-plasticates (by higher diffusion coefficient) (see Fig. 2.12).

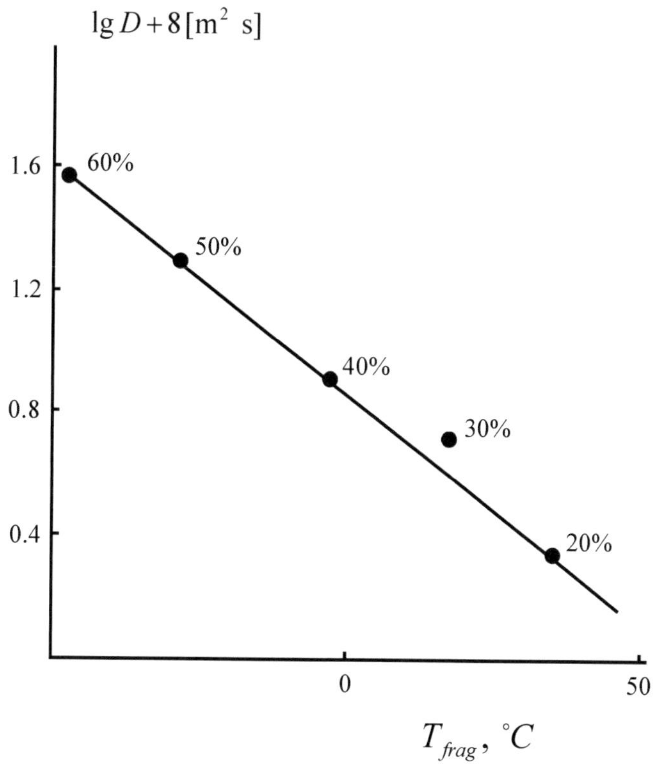

Fig. 2.12. Temperature dependence of friability of PVC-plasticate on diffusion coefficient of tricrezylphosphate.

2.5.2 Time Dependence of Diffusional Desorption of Plasticizers

Early attempts to set quantitative correlation of PVC-plasticate mass loss with aging failed. For example, the calculation performed by Read and Connor that gave 1/4 decrease of dibutylphthalate mass during 500 days of exploitation is disproved by the exploitation practice.

Systematic study of the process of dioctylphthalate diffusional desorption from PVC-plasticate at its contact with different soils are shown in works [98 – 99]. By solving the second Fick equation under the supposition that stationary flow with constant concentration gradient is set since some moment of time, the dependence of plasticizer amount desorbed from covering during time *t* was obtained:

$$m = \frac{Kd\left(12Dt - d^2\right)}{24D} + \frac{4Kd^3}{D\pi^4}\exp\left(-\frac{\pi^2 Dt}{d^2}\right) \text{ (mg/cm}^2\text{)}, \qquad (26)$$

where K is constant value dependent on nature of lowmolecular plasticizer and covering material; D is the diffusion coefficient (mg/dm^2); d is the covering thickness (cm).

The expression obtained describes quite well the kinetics of the desorption process at comparatively long times. The process rate is low under soiling conditions: double-layered PVC covering existed in the soil during 10 years displayed 2 - 3 % of plasticizer (DOP) loss in the upper turned to the soil layer. The layer turned to metal (pipe subjected to insulation) showed no change of plasticizer amount.

The work [100] describes an attempt to disclose the time dependence of PVC-plasticate mass loss, induced by diffusional desorption of plasticizer, also by solving the second Fick equation at border condition $c = 0$. However, as the acceptance D = Const is not correct in this case, the results obtained are in contradiction with the data of exploitation and natural tests. For example, exploitation times until a definite loss of initial resistance to cold is reached differ by a decimal degree.

2.5.3 Changes of PVC-Plasticate Properties during Aging

The change of PVC-plasticate properties at aging is reflected by the Boyer scheme in principle [101] (see Fig. 2.12). Multiple works are devoted to the investigation of changes in electrical, mechanical and thermal properties of PVC-plasticates during exploitation under different conditions [101 – 105].

It was shown in the work 139 that mechanical characteristics and resistance to cold abruptly decrease during exploitation of articles with PVC-plasticate as insulating material (covering) of wires and cables.

The authors of the works [101] limited consideration by phenomenology only, and did not perform the analysis of plasticizer content and changes in PVC.

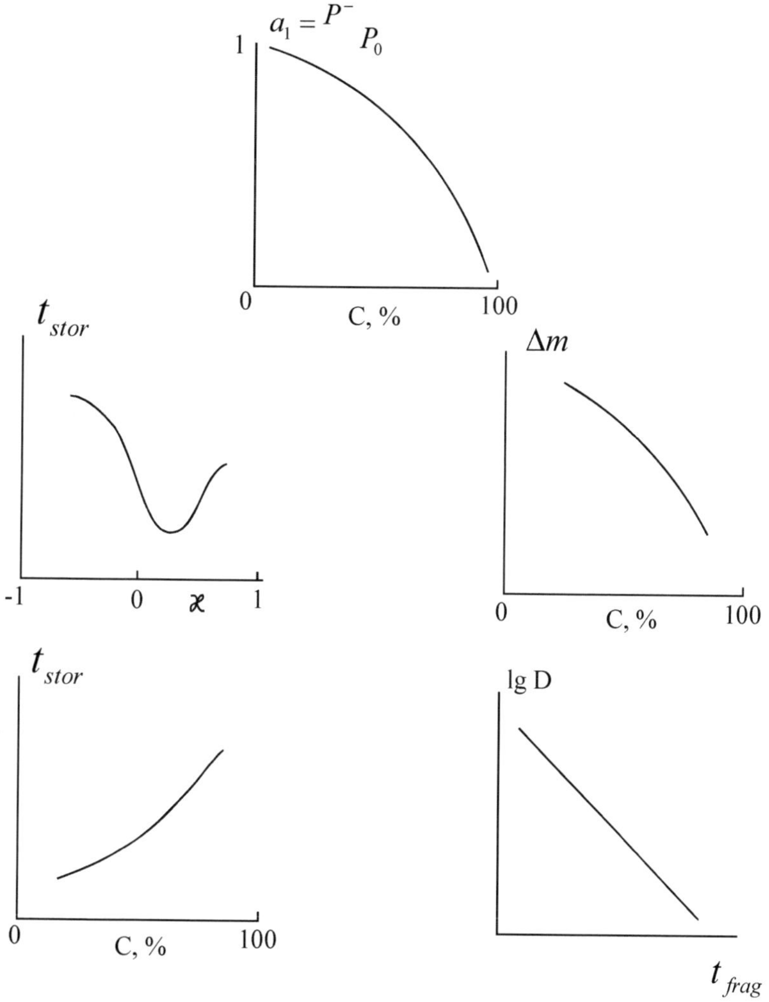

Fig. 2.12. The connection among compatibility, effectiveness and mobility of the plasticizer.

The work [106] stated that at polyethylene contact with PVC-pasticate the process depends on initial concentration of plasticizer in the plasticate. The dependence of migration on plasticizer concentration is of S-form with its linear part in the range of 15 ÷ 35% of plasticizer concentration. Such character of dependence is explained by authors by partial plasticizer solvatation by highly filled polymer. The migration value decreases with the growth of polyethylene crystallinity.

At PVC-plasticate contacts with liquid media extraction of plasticizers is considered, although two processes of interdiffusion proceed in this case.

General view of mathematical description of such a process was given in the work [103, 107]. High effectiveness of low masses was obtained in investigation of extraction of DBP and DOP plasticizers by different oils. In this case larger amounts of oils diffuse into plasticates with DOP plasticizer than with DBP [102]. Similarly to the effect of the diffusion coefficient increase with plasticizer concentration growth the increase of the extraction rate with plasticizer concentration growth is observed.

Similar investigation of food oils applied to extraction of DBP and DOP from PVC-plasticate composition [103] allowed to obtain their effectiveness growth in the sequence as follows: olive oil < corn oil < cotton oil < soy-bean oil. It was found that DOP is extracted faster than DBP.

The following empyric dependence of plasticizer extraction degree was obtained:

$$\alpha^{a} = \sqrt{C_n^{\infty} Mb^{-1}}, \tag{27}$$

here α is the extraction degree; C_n^{∞} is the limit plasticizer concentration; M is plasticizer molecular mass; a and b are coefficients [108].

Benzoresistance of PVC composites 130, resistance to diesel petrol [104] and other media [105] are studied quite well.

The following empiric dependence of vitrification temperature change during PVC-plasticate aging in contact with soils was deduced in the work [98]:

$$T_{vitr} = \frac{RT}{(a - C_0)} \times \frac{Kd\left(12.182\pi^4 d^2 - \pi^4 d^2 + 87.6 d^2 \exp\left(-\frac{\pi^2 Dt}{d^2}\right)\right)}{24.36\pi^4 D m_0}, \tag{28}$$

where a and C_0 are constants.

The process of split formation as the terminal stage of PVC-plasticate aging was considered from positions close to the ideas spoken out by Jurkov. The equation was suggested for the rate of split growth when the material reaches vitrification state:

$$V = \eta A \exp\left(-\frac{B - c\sigma}{RT_{vitr}}\right), \tag{29}$$

where η, A, c are constants; σ is total tension in the sample; T_{vitr} is temperature of the surrounding equal to vitrification temperature.

The time until through splits appear in the covering is determined by the following formula:

$$\tau = \frac{d}{\eta A \exp\left(-\frac{B - c\sigma}{RT_{vitr}}\right)}, \tag{30}$$

where d is the material thickness at which catastrophic growth of destructing splits begins.

It follows from the above mentioned data that the main result of the change of plasticizer concentration in PVC-plasticate during aging is the increase of vitrification temperature and resistance to cold of articles.

2.6 Radiational Aging of Polyvinylchloride

Radiation treatment of polyvinylchloride by ionizing raditiation induces dehydrochlorination accompanied by cross-linking and degradation [109]. At low temperature single-unit double bonds are formed, and at room temperature dehydrochlorination displays chain character and, as a result, conjugated double bonds are formed.

Degradation degree is lower at radiation treatment in vacuum than in air [110]. Radiation-chemical yield of the cross-linking of polyvinylchloride is found higher at polymer radiation in some solvents than in solid state. At radiation in air [110, 111] degradation in non-plasticized polymer proceeds actively than in plasticized one. Apparently, this is connected with the difficulty in polymer radical recombination in more rigid non-plasticized material. Some plasticizers (tricreotylphosphate, phenyl-dicreotylphosphate) display protective effect on polyvinylchloride at radiation treatment.

If unsaturated monomers of complex polyesters are introduced to the recepture, the yield of radiation-chemical linking increases. Adding of tri-allyl ester of cyanuric acid to polyvinylchloride abruptly decreases radiation dose required for obtaining materials with higher thermoresistance and resistance to solvent influence. Radiation doses required for formation of insoluble fraction in polyvinylchloride mixture with triallylcyanurate (the ratio is 90 mass parts to 10 mass parts) decrease from 30 to 0.5 mrad if compared with the initial polyvinylchloride.

Some PVC-plasticates become more rigid under radiation influence, their modulus of elasticity increases [111], whereas relative change of strength at elongation proceeds in the same way as in nonplasticized polymer. The structure of compositions significantly influences their sensitivity to radiation. At radiation treatment of polyvinylchloride plasticized by di-2-ethylhexylphthalate in presence of different stabilizers the stability to break tension and relative elongation decreases monotonously, and at 78 mrad dose it becomes a half of the initial value. In this case gel-fraction content reaches 55%. Sufficient difference in actions of different stabilizers was not found. Tin-containing stabilizer was found more effective filler in presence of $CaCO_3$ comparing with cadmium or barium compounds.

Changes in mechanical and electrical properties of polyvinylchloride plasticates of different receptures were studied in detail in [111].

It was obtained that resistance to break is significantly decreased in the range of 5 ÷ 50 mrad doses, and relative elongation decreases monotonously in all the investigated dose range. More sharp change of resistance was obtained for receptures of plasticates with high concentration of plasticizers. Specific volumetric resistance is abruptly decreased with the growth of the raditiation dose. It is 3 to 37% of the initial in dependence on the recepture compositions at 150 mrad. The resistance to cold decreases abruptly, and ε and tgδ decrease insufficiently. Water absorption by plasticates increases significantly in the range of 1 ÷ 40 mrad doses, which is connected to the formation of polymer groups as a consequence of oxidative processes proceeding under ionizing radiation in presence of air oxygen.

2.7 Photodegradation of Polyvinylchloride

Photoinitiated ageing of poly(vinyl chloride) is a very complex and insufficiently studied process and for this reason it has recently attracted considerable attention by investigators. [112 - 120]. Largely debatable is the role of the reactants known to initiate the photoreactions in PVC, the relative importance of the photo and dark processes responsible for oxidation and polyene buildup, conversion kinetics as a function of light intensity and the composition of environment and the temperature. To a great extent the difficulties encountered in the interpretation of the PVC photodegradation experiments are due to the spatially inhomogeneous kinetics of the process.

The short-wave region of the solar spectrum (below 310nm) makes up almost 5% of the total solar radiation reaching the Earth's surface and features an energy of 358 kJ/mol which is sufficient to break C=C bonds and induce polymer oxidation.

Exposure of VC polymers to light at 250-350 nm develops typical signs of degradation in polymer samples, viz. 1) discoloration from natural to dark-brown or black; 2) surface cracking; 3) brittling or softening of the material; 4) variation of the mechanical properties, i.e. tensile strength, ultimate elongation, impact strength and elasticity modulus; 5) change of transparency; 6) formation of a deposit on the material surface, etc.

What particular photochemical reaction will be the first to occur in the photo-initiated degradation of PVC, depends on the nature of the structural defects in the polymeric chains, the nature of the additives and impurities, including residual initiating agents, protective colloids and the products formed during the processing and service of PVC.

The polymer surface status is of great importance in this respect. Irregularities, porosity, cracks on the polymer surface readily become sites of polymer degradation under UV radiation, especially in the presence of chemically active compounds and sensitizers. All this widens the UV range which is effective in causing PVC degradation to 300-310 nm, 270 280 nm, 253.7-254 nm [121, 122].

PVC has poor light stability in the wavelength range of 253.7 to 310nm, presumably due to the presence of unsaturated C=C bonds, and carbonyl, hydroperoxide and hydroxy groups in polymer chains. The relative activity of one or another chromophore in initiating PVC photodegradation is determined primarily by two factors, i.e. their ability to absorb UV light in the wavelength range under consideration and to participate in the formation of active particles (radicals) which cause degradation of the polymer chains. The alkene unsaturated C=C bonds (both internal and terminal) cannot be the primary initiators of PVC photodegradation under the action of sunlight with $\lambda > 250$nm because they absorb only the UV light at $\lambda < 200$nm. The absorption by conjugated C=C bonds (dienes, trienes, etc.) shifts towards the longer wavelengths, and the more so the longer the polyene [123]. In this < case the extinction coefficient increases considerably, whereby the polyene sequences can effectively absorb the UV radiation even if available in only small percentages in polymeric chains. The initiation of PVC photodegradation may in this situation be due to the scission of certain C—H and C—C bonds, but most probably the C—Cl allyl bonds whose dissociation energy decreases with increasing n [124].

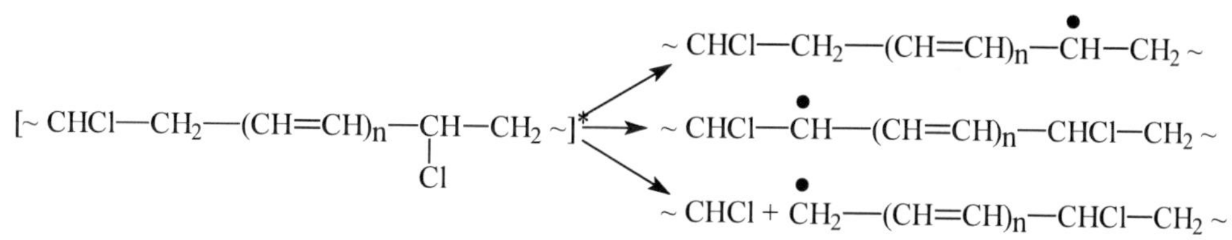

The formation of allyl and alkyl radicals in these reactions has been demonstrated by ESR spectroscopy [113, 123]. The role of unsaturated C=C structures as PVC photodecomposition initiators increases if the polymer sample had previously been subjected to thermal degradation (e.g., during processing) [124].

A prominent part in the initiation of PVC photodegradation is brought about by oxygen-containing groups that are formed during the manufacture of the polymeric product; the first to be named are, of course, the carbonyl structures [112, 113, 120, 125]. Low-molecular ketones are known to be strong PVC photodegradation sensitizers [126], and the photodegradation rate of a vinyl chloride copolymer with carbon oxide directly correlates with the concentration of carbonyl groups. The probable function of carbonyl groups is to transmit the electronic excitation to the VC units of the polymer chain followed by the formation of radicals which degrade the polymer by e.g. the following mechanism [127]:

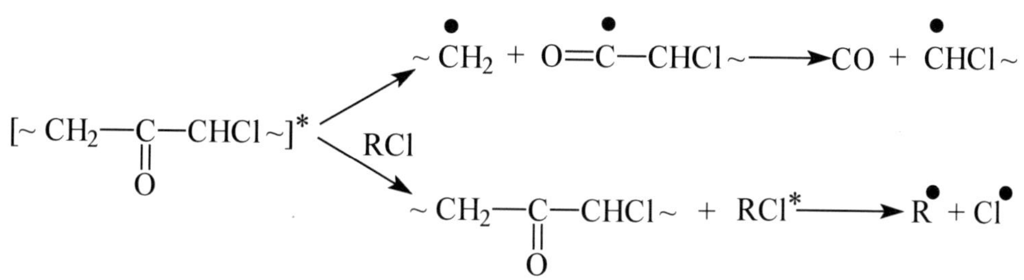

Bearing in mind the important role of the conjugated C=C bonds and carbonyl groups in the polymer's absorption of ultraviolet radiation, the α,β-enone structures

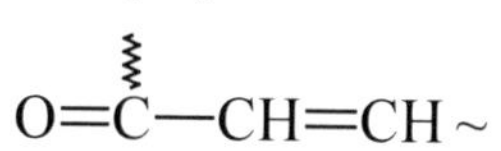

may be assumed to be the most active initiators of PVC photodegradation. Owing to the high extinction coefficients in the wavelength range 250-300 nm, comparable with ε of C=C-polyenes, they may be important light absorbers even if present in small quantities in the polymer chain (ca. 10^{-4} mol/PVC mol).

Other potential initiators are peroxides and hydroperoxides in PVC [115] which undergo homolysis under UV radiation and thus initiate polymer degradation [128, 129].

On the whole, carbonyl allyl groups are evidently the most important initiating chromophores occurring in PVC, since there are normally no peroxides or hydroperoxides in commercial polymeric products.

In studying light ageing [96] we assumed that degradation processes occur in a surface layer and depend on the spectral composition of incident radiation. PVC films of different thickness prepared by slow solvent (dichloroethane) evaporation from 5% polymer solution were illuminated by light with $\lambda = 254$ nm and $\lambda > 300$ nm. PVC oxidation was monitored by the buildup of carbonyl groups detect by absorption IR spectroscopy at 1720 cm^{-1}, and PVC dehydrochlorination was studied following the formation of polyenes detected from the UV and visible absorption spectra.

As seen from the kinetic data on illumination with light of $\lambda = 254$ nm, the oxidation degree is independent of the sample thickness, which suggests that light-induced oxidation processes take place only within a very thin subsurface PVC layer, less than 20 μm in thickness, whereas the dehydrochlorination extent is highly dependent on the sample thickness, because of degradation reactions in the bulk.

The PVC ageing kinetics in samples illuminated with milder UV light ($\lambda > 300$ nm) is qualitatively different: both the oxidized and dehydrochlorinated fractions depend on the sample thickness, but samples are aged inhomogeneously. Hence, under these conditions, PVC is oxidized in deep polymer layers rather than only near the surface. The kinetics of PVC dehydrochlorination in samples illuminated at $\lambda > 300$ nm and $\lambda = 254$ nm differs only quantitatively and also by the lack of a pronounced induction period in samples illuminated at $\lambda > 300$ nm. Another important specific property of PVC revealed experimentally is the dependence of its absorption spectrum on the spectral composition of illuminating light, which is presumably associated with different polyenes formed upon illumination with light from different sources.

Thus investigation of the spatial distribution of photodegradation products and its variation with time in polymer may he a sensitive method for the study of the mechanisms of polymer photodegradation and photostabilization [7, 96, 115, 121, 130 – 134].

3 EXPERIMENTAL

3.1 Investigation Methodics and Technique

3.1.1 Thermoanalytical Technique

Thermoanalyzers 1090 'Dupont Instruments' (USA) and TA 3000 'Mettler' (Switzerland) were used for the performance of thermal analysises. First of them allows to carry out investigations in 73 ÷ 1473 K temperature range, and the second up to 1273 K only. Measurements are performed in isothermic regime (± 1 K measurement mistake), and dynamic regime varying heating rate from 0.1 to 100 K/min.

Thermoanalyzers are designed for application with different physical modulus:

- Thermogravimetric analyzer performing in isothermic and dynamic regimes recording the mass loss derivative curve (TG-951 'Dupont' and TG-50 'Mettler'). The mass of specimen is 25 ÷ 30 mg. Usual weighting mistake is $\pm 0.5\times10^{-3}$ mg. Investigation were performed in air with differentiating scanning calorimeter (DSC-910 'Dupont' and DSC-30 'Mettler'). The mass of samples was 10 ÷ 15 mg, accumulated in aluminum caps. Quantitative measurements were performed after preliminary leveling by standards (sapphire and indium).

- Thermomechanical analyzer (TMA-943 'Dupont' and TMA-40 'Mettler'). Sample mass is 10 ÷ 20 mg. Usual error of the size determination is $\pm$ 1.0×10^{-4} mm. Investigations were performed at 20 g load and 10 K/min rate of temperature increase.
- Dynamic mechanical analyzer (DMA-982 'Dupont'). The investigation was carried out on samples of standard (3.0 ± 0.1)×(6.0 ± 0.1) mm size and (1.0 ± 0.1) mm thickness.

3.1.2 Spectral Technique

Fourrier IR-spectrometer 'Qualimatic' Digilab (USA) was used for obtaining IR-spectra. This appliance represents a single-beam spectrometer with Michelson interpherolyte. Spectra were recorded in 4400 ÷ 400 cm^{-1} range, maximum resolution was ± 0.1 cm^{-1}; argon was used as the working gas phase (pressure 2 atm., expenditure rate 600 l/hour). IR-spectra of penetration and reflection were obtained. The accessory of multiple broken complete internal reflection with CP-5 cristal was used in the latter case. Sample sizes were (2.0 ± 0.1)×(5.0 ± 0.1) mm, the film thickness was 0.1 ÷ 0.35 cm.

An infrared microspectrometer was applied for obtaining IR-spectra (of reflection and penetration) of microsamples possessing squares up to 20 μm^2.

UV-spectra of samples were investigated on UV-spectrometer SP-800 Pye Unicam (UK) in 185 ÷ 700 nm range.

3.1.3 Chromatographic Analysis Technique

Chromatographic analysis was performed on gas-liquid chromatograph of 'Light' series. Catarometer was used as a detector, gelium was used as the gas-carrier. A column 1000 mm long and 3 mm in diameter, filled by Chromaton (particle diameter is 25 ÷ 32 μm) nuorished by Apieson L liquid phase, was used at the analysis of plasticizers. Temperature of the column was 563 K, evaporator and detector - 623 and 573 K, respectively; current in detector was 80 A. Plasticizer solution in diethylic ester (specimen size was 50 μm) was introduced directly into the evaporator by a microinjector.

Fluid high-pressure chromatography was performed on fluid chromatograph 880 Dupont in the regime of adsorptional and gel-penetrating (sieving) chromatography. Zorbax and SE-100 columns were applied, respectively, in connection with SE-500 column. Moving phase was presented by tetrafuran of optical purity, specimen volume was 100 μl possessing 0.4% concentration; the detector used was differential refractometer with 0.2 sensitivity; ultra-violent spectrometer (wavelength from 200 nm to visible range) and infra-red spectrometer (wavelength from 2.15 to 15 μm) were also used.

3.1.4 Mechanical Technique

Mechanical properties were investigated on universal test-machine Instron-1122 (UK). Maximal load was 5 KII, minimal range of applied load was 0 ÷ 0.1 II. The rate of set deformation was 0.05 ÷ 1000 mm/min, usual measurement mistake was ± 0.1%.

Microporous structure of samples was investigated on porometer Autopore 9200 Cultronics (France). The apparatus works in automatic regime and allows to measure the total volume and surface square of pores, volumetric and specific density, and pore distribution by sizes. Mercury is used as the penetrating liquid. Pores sized from 200 to 0.003 μm are measured in the range of applied pressures from 2 to 4200 kg/cm^2. Pore radius is determined from the following expression:

$$r = \frac{2\sigma \cos\Theta}{P}, \tag{31}$$

where r is radius of a cylindric pore; σ is mercury surface tension (480 dyn/cm); Θ is contact angle (~ 140°); P is applied pressure. Sample volume is 2 ÷ 5 cm^3, resolution by mercury volume is 5×10^{-4} cm^3, common mistake at pressure measurement is ± 0.5%.

Centrifuge T-24 Intermed (Germany) was used for determination of adhesion forces. Maximal rate is 16,000 rot/min.

Linear sizes during aging were determined with the accuracy ± 0.1 mm on samples possessing initial sizes of (20.0 ± 0.1) ×(100.0 ± 0.1) mm. The thicknesses were measured by micrometer with common mistake of measurement ± 0.01 mm.

Lead salt concentration in samples was measured by atomic-absorptional spectrophotometer 1000 Perkin Elmer (USA) applying acethylene-air mixture. Minimal concentration was 1×10^{-4} mmol/l. Calibration was performed by three standard solutions.

3.2 Fundamentals of PVC-plasticate Investigations

During storage and exploitation PVC-plasticate articles are subjected to influence of temperature, humidity, air oxygen. They are also treated by mechanical loads, aggressive media and their vapours, ozone, ionizing radiation and physical fields during exploitation (see Fig. 3.1).

Factors of aging	Physical and chemical properties	Exploitation properties
Temparature Oxygen Humidity	Diffusional desorption of plasticizers and stabilizers	Mechanical and dielectric properties
Mechanical loads	Dehydrochlorination	Colour
Aggressive media	Thermoxidative degradation	Sorption-diffusional properties
Ionizing radiations	Water sorption	

Fig. 3.1. The change of PVC-plasticate relative mass at heating at 10 K/min rate. Concentration of dioctylphthalate: curve 1-10; curve 2-20; curve 3-30; curve 4-40; curve 5-50; curve 6-60 mass parts per 100 mass parts of PVC, respectively. Inset: the dependence of the mass loss on dioctylphthalate concentration at heating to 400°C.

The analysis of data from literature shows that the following processes, which lead to aging of the material, i.e. the decrease of its exploitational properties, proceed in PVC-plasticate under influence of the above mentioned factors (see Fig. 3.1):

- Dehydrochlorination leads to HCl extraction and formation of polyene sequences in PVC macromolecules. Quantity and length of polyene sequences define color of the material from light-brown to black;
- Thermoxidative degradation induces the decrease of molecular mass and formation of carbonyl-containing groups;
- Diffusional plasticizer desorption leads to the increase of vitrification temperature of the material;
- Processes of sorption of water and other low-molecular compounds by the material from the surrounding.

Chemical and physical processes proceed simultaneously during PVC-plasticate aging, which change chemical composition and structure of the material. Not all processes can proceed with noticeable rate under real conditions of storage and exploitation. And moreover, proceeding of a part of them will cause the change of the exploitation properties. That is why two tasks appear at the investigation of every polymer material aging, PVC-plasticate in particular. It is necessary:

1. To disclose processes proceeding in polymer material during aging, responsible for the change of the exploitation properties at application of the present article;
2. To obtain formal-kinetic equations which connect indexes of exploitational property with time and factors of aging.

The unity of these kinetic data are called the model of aging. In this case it should be taken into account that these data, as a rule, allow to make no conclusion about general mechanism of aging of the present polymer material, but it is enough information for description of kinetics of the polymer material aging under current exploitation conditions.

The question is: how to control kinetics of processes proceeding in polymer materials at aging under induced and natural conditions.

Characteristic indexes of aging are used for quantitative estimation of polymer material properties at aging. They are the parameters, which highly characterize exploitation suitability of the material in article. Indexes of the exploitation properties of the material, indirectly connected to the exploitation properties, are most often chosen as characteristic indexes of aging. They are, for example, oxygen absorption at thermoxidative degradation, HCl extraction in dehydrochlorination, material mass decrease at plasticizer desorption, etc.

Characteristic indexes of aging must:

- Uniformally characterize the present process proceeding during aging;
- Be easily registered;
- Change when the material in article is yet suitable for exploitation.

The following indexes are used as characteristic indexes of aging in order to differ it from changing of properties, induced by the change of environment conditions (temperature, humidity):

- Index of fundamental material property (chemical composition, molecular-mass distribution of polymer, physical structure of the material); indexes of fundamental properties do not depend on environment conditions;
- Index of a property, measured under fixed conditions.

The analysis of data from literature shows that the main process in PVC-plasticate at aging at room temperatures during exploitation is, probably, the diffusional desorption of plasticizer. At temperatures over 60°C proceeding of thermoxidative degradation is also probable, and it may be accompanied by dehydrochlorination. It is advisable to search the proceeding of the first process by the amount of plasticizer remaining in the PVC-plasticate. Existence of thermoxidative degradation and dehydrochlorination can be identified by the change of the molecular-mass distribution in polymer and accumulation of double bonds in it, respectively.

3.3 Investigations of Kinetics of PVC-plasticate Aging

3.3.1 Subjects of Experimentation

Investigations were performed on insulation of BWPV wires, exploited on articles '**I**' and '**II**' during long time (maximal exploitation time was 28 years), and insulation of wires of analogous trade marks, which were not exploited and were received from industrial processing not later than a year before. Moreover, PVC-plasticate films of the following recepture 0.1 ÷ 0.35 cm thick were also studied: PVC C-63M - 100 mass parts, DOP - 24 mass parts, cadmium stearate - 3 mass parts, and specially prepared samples with different dioctylphthalate concentrations.

3.3.2 Investigations of Plasticizer Diffusional Desorption

Plasticizer yield from PVC-plasticate is a two-stage process that includes:

- Plasticizer diffusion to article surface;
- Plasticizer desorption from article surface into the surrounding.

As plasticizer concentration in the material decreases during aging, it is wrong to use the second Fick equation for description of the diffusion process. In this case empyric equations are used, for example:

$$m_t = Kt^n, \quad (32)$$

where m_t is the plasticizer amount, extracted during time t; n and K are the parameters (n changes in the range $0.5 \leq n \leq 1$).

Plasticizer concentration on the material-air interface is determined according to the mass transfer laws, the simpliest of which is the following:

$$m_t = \beta S\left(\bar{C}_0 - C_{air}\right)t, \tag{33}$$

where β is the mass transfer coefficient.

The limiting stage of the process can be determined by m_t dependence on article thickness.

Mathematical treatment of methods of plasticizer amount determination in PVC-plasticate

The following techniques were used for determination of plasticizer amount in PVC-plasticate: thermogravimetric technique, chromatographic technique and UV-spectrophotometric technique.

Thermogravimetric technique

Heating of PVC-plasticate samples (25 ÷ 30 mg) to 400°C causes the decrease of sample mass, induced by plasticizer desorption as well as by PVC dehydrochlorination. Fig. 3.2 shows curves of the mass change of plasticate samples with different plasticizer concentration during heating at the rate of 10 K/min. The mass decrease practically stops at 400°C, and it is possible to calculate plasticizer content in plasticate by total mass loss. Fig. 3.2 also shows the linear dependence between remaining mass $\frac{m_0 - m_{400°}}{m_0}$ and dioctylphthalate amount in the initial material.

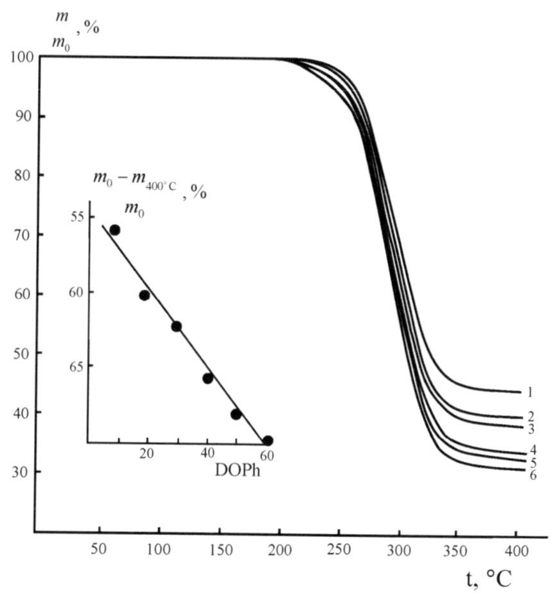

Fig. 3.2. Chromatogram of the mixture of plasticizer and leveling dependence. DDP = diisodecylphthalate, DOP = dioctylphthalate, and DOC = dioctylcebacynate.

Experiments showed that if PVC-plasticate sample is exposed at 300°C during 25 min, we can reliably calculate the remaining mass of the sample after plasticizer desorption and PVC dehydrochlorination.

Chromatographic technique

Extraction of plasticizers by diethyl ether was performed at 45°C during 6 hours. Solution of plasticizers in diethyl ether was analyzed on gas-fluid chromatograph.

Fig. 3.3 shows typical chromatogram of three plasticizers and leveling lines, which connect peak squares with plasticizer amounts in injected probes.

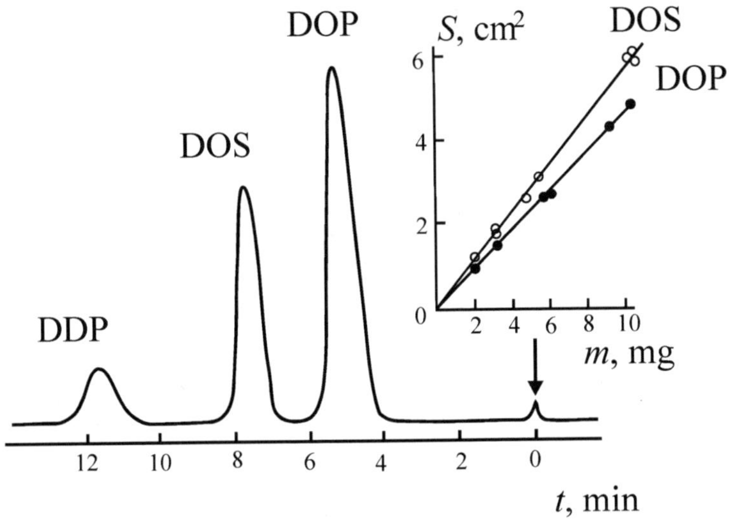

Fig. 3.3. UV spectra of dioctylphthalate in tetrahydrofuran.

The method allows analysis of mixtures of plasticizers, however it is impossible to extract plasticizers completely under present conditions after extraction by diethyl ether (the remaining amount is 5 ÷ 6%).

UV-spectrophotometric technique

PVC-plasticate specimen was dissolved in optically pure tetrahydrofuran at 80°C during 1 hour. The solution was filtered by 'Millipore' filter with average pore size of 0.45 μm. It is possible to use Russian lavsan-based filter with diameter of pores 0.40 ± 0.05 μm (produced by International Institute of Nuclear Researches, Dubna, Russia).

Dioctylphthalate solution in tetrahydrofuran possesses two absorption bands at 230 and 274 nm (see Fig. 3.4). Absorption coefficients obtained from leveling experiments equal 19.3 ± 0.2 and 3.2 ± 0.1 l/mg×cm, respectively. Plasticizer concentration is calculated by the following formula:

$$C = \frac{DV}{\varepsilon ml} \times 100, \tag{34}$$

where D is the optical density; ε is the absorption coefficient; V is the solution volume; m is the mass of the probe analyzed; l is the optical cuvette thickness.

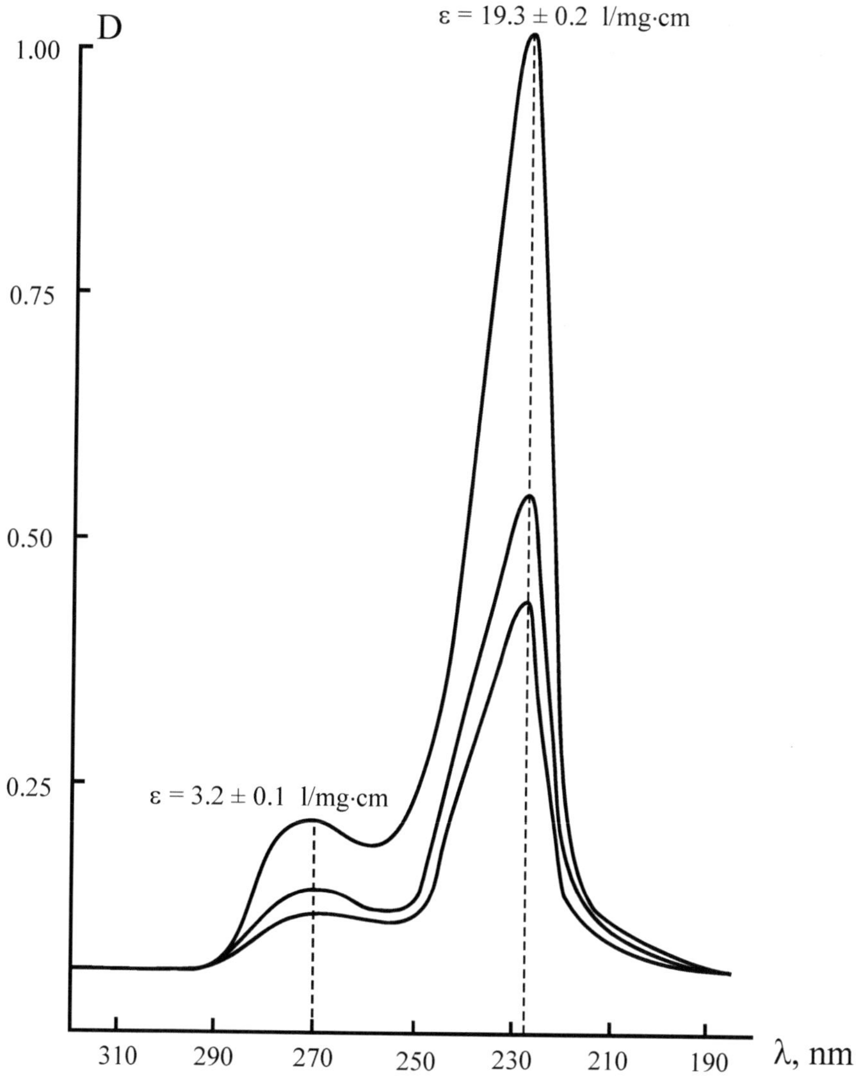

Fig. 3.4. The dependence of ln (m_o/m) on ln t for diffusional desorption of dioctylphthalate from PVC-plasticate at 135°C (curve 1) and 120°C (curve 2).

The method allows analysis of the concentration of a single or mixed (several) plasticizers in PVC-plasticate obtaining 1 ÷ 2% accuracy.

Formal-kinetic description of the plasticizer diffusional desorption by change of sample mass

Kinetic curves of diffusional desorption of plasticizers in 20 ÷ 165°C temperature range are satisfactorily described by the following empiric equation:

$$\ln\frac{m_0}{m_t} = Kt^{0.62\pm0.03}, \tag{35}$$

where K is the desorption constant.

Fig. 3.5 shows typical treatment of experimental data on dioctylphthalate desorption from PVC-plasticate at 120 and 135°C; the data were obtained by the thermogravimetric technique.

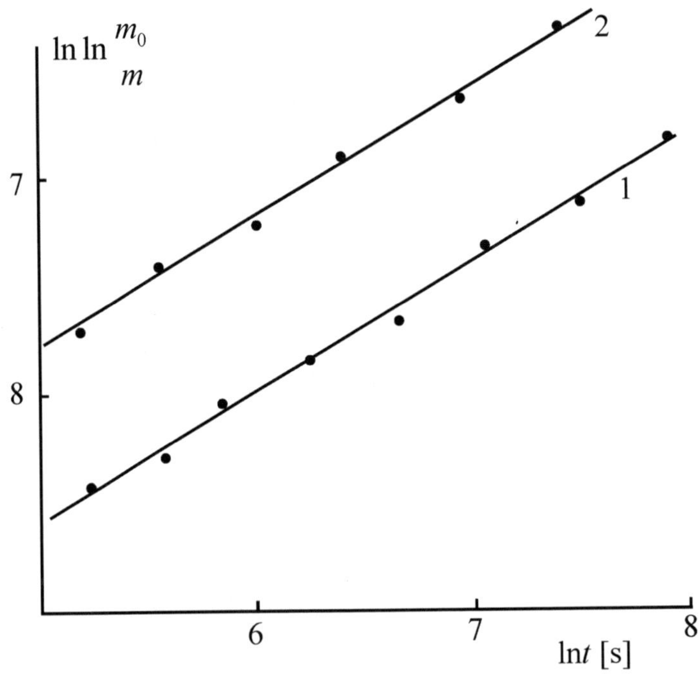

Fig. 3.5. Kinetic curves of dioctylphthalate desorption from PVC-plasticate at temperatures of 120°C (curve 1), 135°C (curve 2), 150°C (curve 3), and 165°C (curve 4).

The equation (35) describes experimental data obtained by the UV-spectrophotometric technique. However the data obtained by the chromatographic technique (see Fig. 3.6) are not described by this equation. It is, probably, connected with insignificant occuracy of the method.

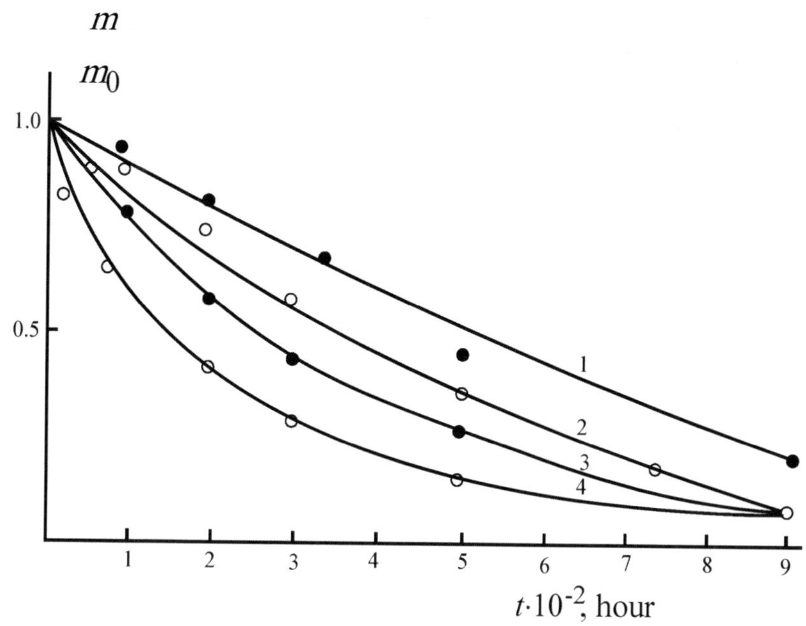

Fig. 3.6. Generalized kinetic dependence of the diffusional desorption of dioctylphthalate from PVC-plasticate in In $(m_o/m_t) - Kt^{0.62}$ coordinates: (●) 135°C; (O) 120°C; (Δ) 105°C; (□) 90°C; (X) 47°C.

Fig. 3.7 shows generalized dependence of the diffusion desorption of dioctylphthalate from PVC-plasticate in $\ln\frac{m_0}{m_t} - Kt^{0.62}$ coordinates. It shows that the process, probably, proceeds by one and the same mechanism in the whole temperature range.

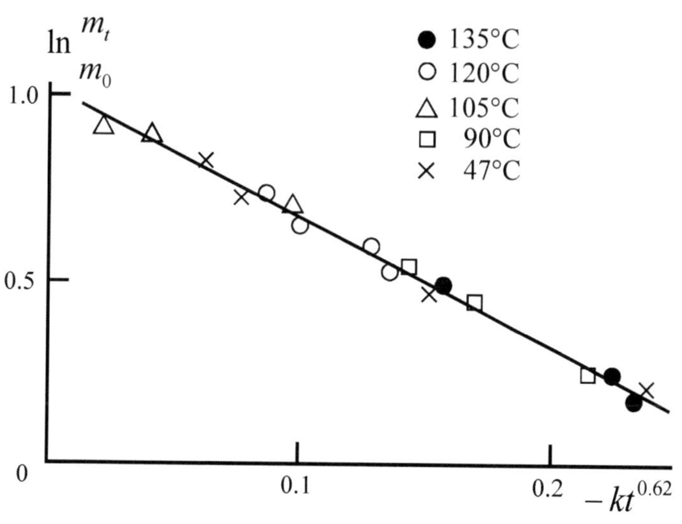

Fig. 3.7. The change of mass and linear sizes of PVC-plasticate thin film upon heating in air up to 150°C.

Changes of linear sizes of articles made from PVC-plasticate

We should naturally expect the change of linear sizes of PVC-plasticate articles at the diffusion desorption of plasticizer from PVC-plasticate.

Linear and volumetric shrinkage of thin films were determined according to the following formulas:

$$\Delta l = \frac{l_0 - l_t}{l_0} \times 100, \tag{36}$$

$$\Delta V = \frac{V_0 - V_t}{V_0} \times 100, \tag{37}$$

where l_0 and V_0 are initial values; l_t and V_t are values after aging during time t.

Shrinkage of wire insulation by length and diameter proceeded according to formulae (36) and (37) and one more formula that follows:

$$\Delta d = \frac{d_0 - d_t}{d_0} \times 100, \tag{38}$$

where d_0 and d_t are initial and current values, respectively.

Samples possessing sizes of 100×20 mm were investigated in experiments with thin films.

Fig. 3.8 shows changes of mass and linear sizes of PVC-plasticate thin films at heating in air at 150°C. Total change of mass is (27.0 ± 0.2)% at kinetic curve layout to the plateau. The mass change is described by the equation (35), the rate constant equals 2×10^{-13} $s^{-5/8}$. Linear size change is satisfactorily described by the order one equation as follows:

$$\Delta l = \Delta l_\infty\left[1 - \exp(-Kt)\right]. \tag{39}$$

The value of Δl_∞ is found (17.0 ± 0.3)% by length and (7.0 ± 0.4)% by width, respectively. The value of the rate constant is found (5.5 ± 0.2)×10^{-6} s^{-1}. The limit volume change of the PVC-plasticate film is (23.0 ± 0.4)%.

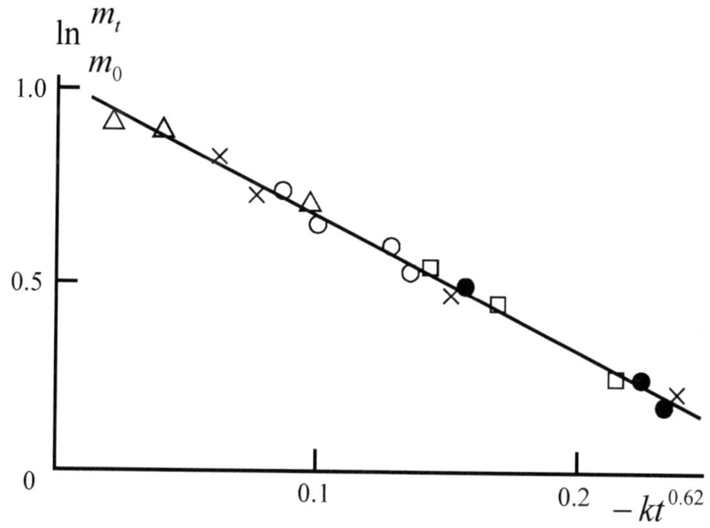

Fig. 3.8. The dependence of Δm on Δl (length) of PVC-plasticate upon heating in air at 150°C.

It should be mentioned that until the mass change reaches ~25% the linear dependence between Δm and ΔV takes place, and after this point the break-shold of the above mentioned dependence occurs (see Fig. 3.9).

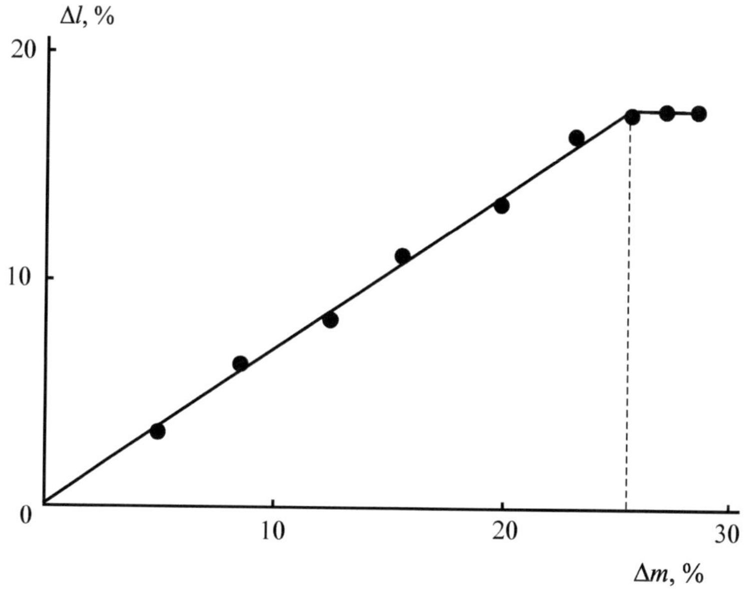

Fig. 3.9. The change of length and diameter of wire isolation (sample length is 100 mm) upon heating in air at 150°C.

The change of linear sizes of wire insulation (100 mm sample length) proceeds in less degree than that for PVC-plasticate films under the same aging conditions. At the same mass change ((27.0 ± 0.2)%) the change of length is found (9.0 ± 0.2)%, and (3.8 ± 0.2)% of diameter (see Fig. 3.10). Lower shrinkage value by length of wire insulation is explained by the existence of adhesion forces of polymer to metal, which prevent wire insulation shrinkage.

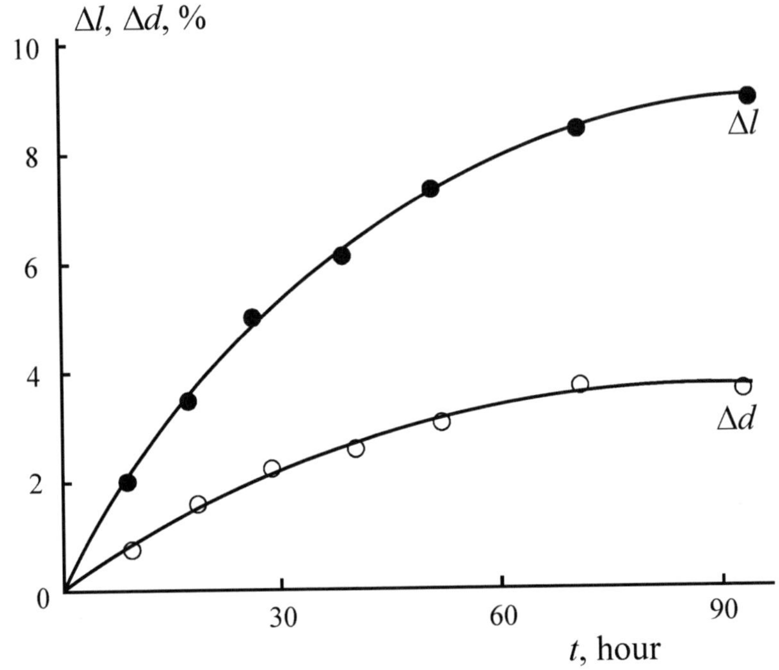

Fig. 3.10. Initial rates of dehydrochlorination measured by accumulation of polyene fragments of PVC-plasticate at different temperatures.

3.3.3 Investigations of Thermooxidative Degradation

The analysis of literature data shows that thermooxidative degradation of polyvinylchloride is accompanied by dehydrochlorination. In such a manner it is advisable to control thermooxidative degradation by the change of molecular-mass distribution of polyvinylchloride and by double bond accumulation in it.

Molecular-mass distribution of polyvinylchloride

Molecular-mass distribution (MMD) of polyvinylchloride was determined by gel-penetrating chromatography technique. Both initial and aging-treated PVC samples possessed bimodal distribution with peaks, maximums of which are corresponded to hold-back times of (4.72 ± 0.03) and (7.96 ± 0.02) min. Corresponding values of the mean-viscous molecular mass of PVC are 4.8×10^4 (A) and 3.6×10^3 (B), respectively.

Table 3.1 shows ratios of PVC MMD components at aging at different tempreature conditions.

Table 3.1. Correlation of PVC MMD components in different tempreature-time regimes of aging.

Time of aging, hour	**Temperature,** °C	**Parts of PVC MMD components, %**	
		A	B
0	20	70.8	29.2
4.4×10^4	20	70.6	29.4
100	70	70.5	29.5
50	80	70.4	29.6
100	100	72.5	27.5
100	110	74.9	25.1
100	120	88.0	12.1
3.5	140	97.3	—

Concentration of high-molecular component (A) increases monotonously at aging at 140°C, and the peak of lowmolecular component (B) completely disappears 2.5 hours after aging termination.

There was no sufficient change of MMD (> 2%) observed in PVC-plasticate samples exploited under soft conditions during 20-28 years.

In such a manner we can assume that exploitation of PVC-plasticate-made articles under soft conditions during 28 years cause no noticeable changes in polymer MMD. At higher temperatures (> 100°C) the aging during 100 hours causes significant changes of MMD to the side of high-molecular component concentration growth. It may be, probably, explained by cross-linking of low-molecular PVC fragments.

Dehydrochlorination

Concentration of polyene fragments, formed in dehydrochlorination, was measured by fluid chromatography technique. Zorbax ODS column 250×4.6 mm size was used; the flow rate was 1 ml/min; UV-detector with 333 nm wave-length was used, which sensitivity was 0,02. Tetrahydrofurane was used as the moving phase; probe volume was 50 μl, concentration of dissolved polymer was 0.05%.

The initial polymer possessed the peak of (2.12 ± 0.01) min hold-back time. A new peak, possessing (1.76 ± 0.01) min hold-back time, appears at aging. This peak is corresponded to polyene fragments with five and more double bonds. Table 3.2 shows parts of peaks with hold-back times of (2.12 ± 0.01) min (A) and (1.76 ± 0.01) min (B), respectively, for different conditions of aging.

Table 3.2. Ratios of peaks of initial and polyene fragment-containing PVC for different temperature-time regimes of aging.

Aging duration, hour	Temperature, °C	PVC-plasticate Trade mark	Parts of peaks, %	
			A	B
0		I-40-13	100	0
1×10^4	70	I-40-13	97.2	2.8
2.2×10^5	17	I-40-13	96.4	3.6
0		I-40-12	100	0
1×10^4	70	I-40-12	97.6	2.4
1.9×10^5	17	I-40-12	97.2	2.8
0		S-70	100	0
1×10^4	70	S-70	97.8	2.2
1.9×10^5	17	S-70	97.6	2.4

Initial rates of dehydrochlorination were measured by accumulation of polyene fragments at 120 ÷ 180°C temperature range (see Fig. 3.11).

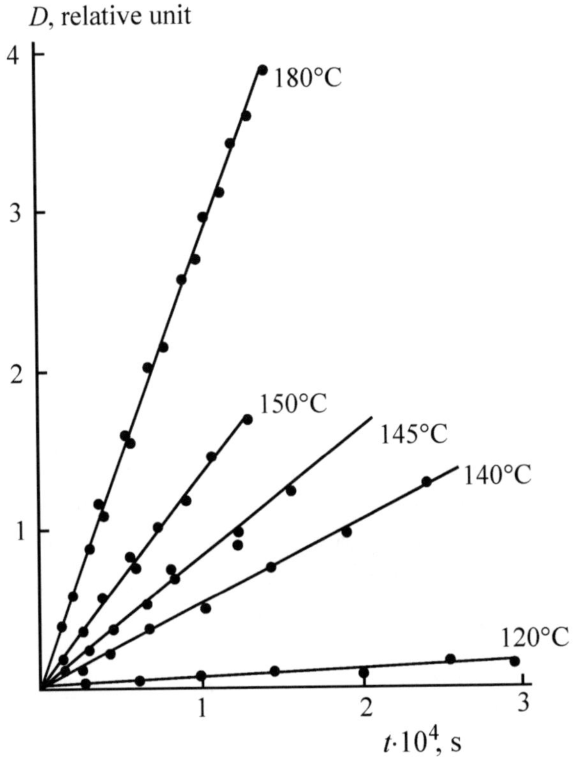

Fig. 3.11. Typical methods of braid protection: (A) twisting; (B) pipe application.

The value of effective activation energy is found 113 ± 5 kJ/mol, that much better correlates with the activation energy of catalytic dehydrochlorination than with the activation energy of dehydrochlorination in the presence of oxygen (see Section 2.2).

3.3.4 Investigations of Microporous Structure

Microporous structure of samples of the following sequence of I-40-13 trade mark PVC-plasticate was investigated by mercury porometry technique: the initial sample produced in 1985; the sample treated by aging in air at 70°C during 8949 hours; the sample dismounted from the article '**II**', produced in 1955 and exploited during 5296 hours, which is not in use at present. Mean-equivalent temperature of exploitation was ~20°C.

Main characteristics of the microstructure are shown in the Table 3.3.

Table 3.3

Characteristics	Initial sample	Aging at 70°C, 8949 hours	Natural aging at 20°C, 30 years
1. Total pore volume, cm^3/g	0.172	0.0381	0.0656
2. Total pore surface, m^2/g	4.54	7.64	24.1
3. Average pore diameter, μm	0.015	0.020	0.011
4. Capillarity	4.6	18.1	6.2

It is seen that a sufficient increase of the total pore volume occurs during aging. It should be mentioned that although average pore diameter does not practically change, the pore distribution by sizes changes significantly. For example, if pore content in the initial sample is several percents, the sample treated by 30 year aging displays 20% of pores, and pores possessing 2 ÷ 3 μm radius appear in it.

4 APPLICATION

4.1 Characteristics of Board Wires (BW) with PVC-plasticate Insulation and Application Conditions

BWPV wires are exploited as a part of board electric network (BEN) of articles '**I**' and '**II**'. BWPV is a board wire with insulation made from polyvinylchloride plasticate covered by brainding from antisepted cotton thread, covered, in its turn, by antisepted nitrocellulose vanish. Wires are designed for mounting BEN under conditions of the environment temperature from -60 to +70°C for voltage below 220 V. Contrary to industrial wires, aviation electric wires must reliably work in a wide range of temperatures, pressures, mechanical loads, in aggressive media and their vapours. Exploitation factors which treat wires are shown in Table 4.1.

Table 4.1. Exploitation factors treating BEN wires of articles 'I' and 'II'.

Type of treatment	Level of influence
1. Environment temperature, °C	From -60 to +70°C
2. Relative humidity, %	98 (at 20°C)
3. Rarefaction , mm Hg	To 140
4. Net voltage, V - direct current -alternate current	 Up to 30 Up to 115
5. Alternate current frequency, Hz	400

Wire insulation is also treated by air oxygen and ozone influence during exploitation. As BEN wires are mounted inside planner of an article, there is no influence of ultra-violet radiation on them.

Heat influence significantly differs by its intensity in dependence on the disposition of wire cord. Such zones are conditionally divided into 'hot' (possessing temperatures over +60°C) and 'normal' (with temperatures below +60°C). Engine zones, the zones of hot air pipelines of anti-icing system and some other aggregates of the article are devoted to 'hot' zones. In other zones heat influence is in the range of the surrounding air temperature.

The specific feature of BEN wire exploitation is the difference in heat treatment on land and during flight. For example, in summer time in regions with hot climate insulation is treated by temperatures up to +50°C for long time, and then it is cooled down to -50°C in upper atmosphere during flight.

Exploitation conditions define the following features of an aviation cable construction:

- Current-conducting vein is twisted from a multiple of thin wires covered by protecting covering. This provides cable flexibility, vibroresistance and protects material of the current-conducting vein from chemical reactions, proceeding between insulation covering and vein;
- Insulation and protective covering of cables is prepared from highly resistant polymer materials and fibres stable to high temperatures, mechanical and chemical influences.

The following subdivision of aviation electric cables according to their application is used:

- Board cables applied to repairing external (interblock) parts of BEN;
- Mounting cables applied to mounting BEN parts in panels, electric panels and blocks;
- Special cables (high-voltage, compensation, etc.).

Aviation electric cables are also subdivided according to materials used in them:

- Heat-resistant, working temperature over +70°C;
- Heat-nonresistant (usual), working temperature below +70°C.

The present review considers data of investigations of BWPV and BWPV-E cables (TC 16-06-370-69). These cables are designed for BEN mounting in conditions with the surrounding temperature in -60 to +70°C range at voltage below 220 V.

BWPV is the board cable with insulation made from polyvinylchloride plasticate covered by braiding of antisepted cotton thread, which, in its turn, is covered by antisepted nitrocellulose vanish. BWPV-E cables possess additional screen braiding made from tinned copper wires. Cables are twisted into braids before mounting on articles. There are several methods of cable braid protection from influences of high temperatures, aggressive media, mechanical damages which depend on exploitation conditions:

- Cable twisting;
- Covering by protective belts, coverings and glueing tissues;
- Putting pipes on cables;
- Applying screen braids on cables.

Typical methods of braid protection are shown in the Fig. 4.1.

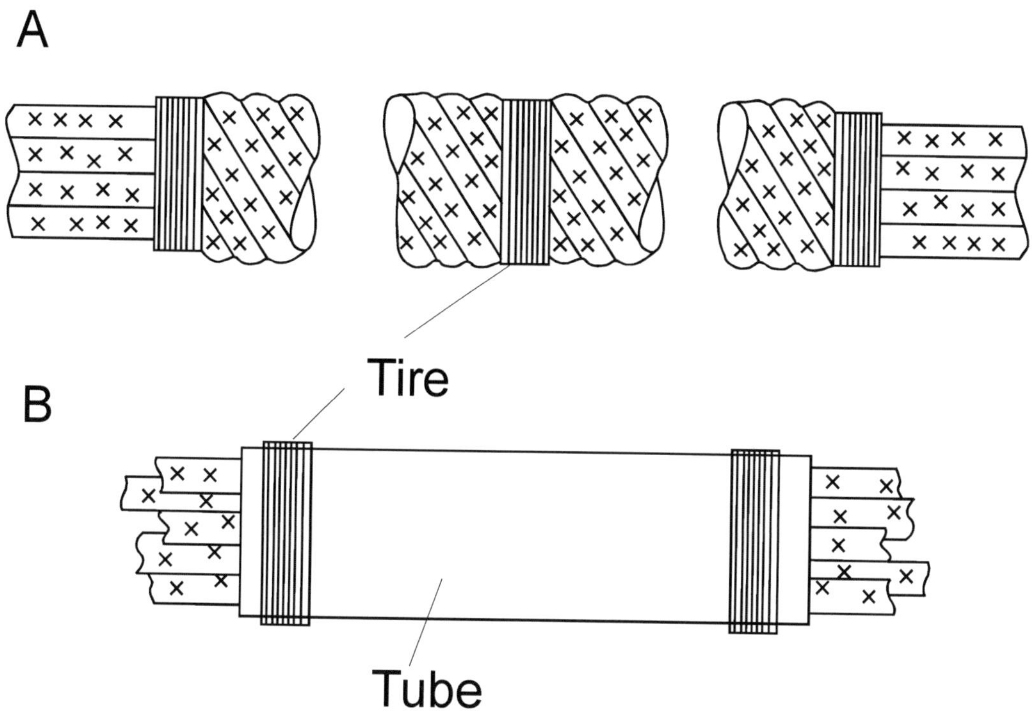

Fig. 4.1. Typical blocks of braid mounting on construction elements of an aircraft: (top) fastenings to stringers, ribs, and so forth; (bottom) mounting of a moving braid part (braid connection with landing gear and fuselage).

Braid mounting on board a flying facility is performed with the help of normalized clips: collars, clamps, etc. Typical blocks of applying braids and cables to construction elements of aircraft are shown in the Fig. 4.2.

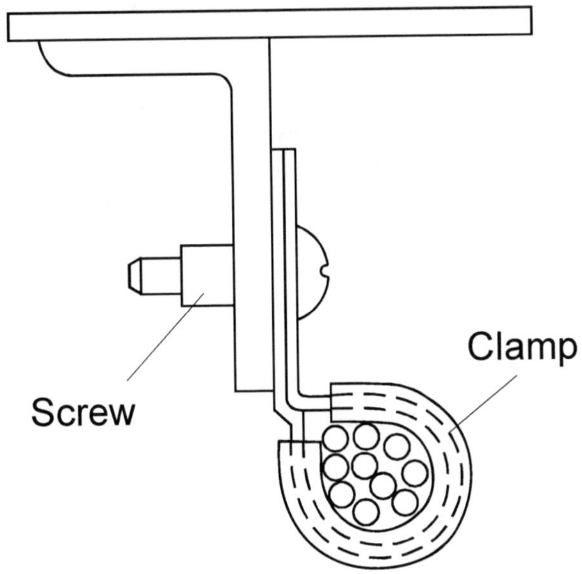

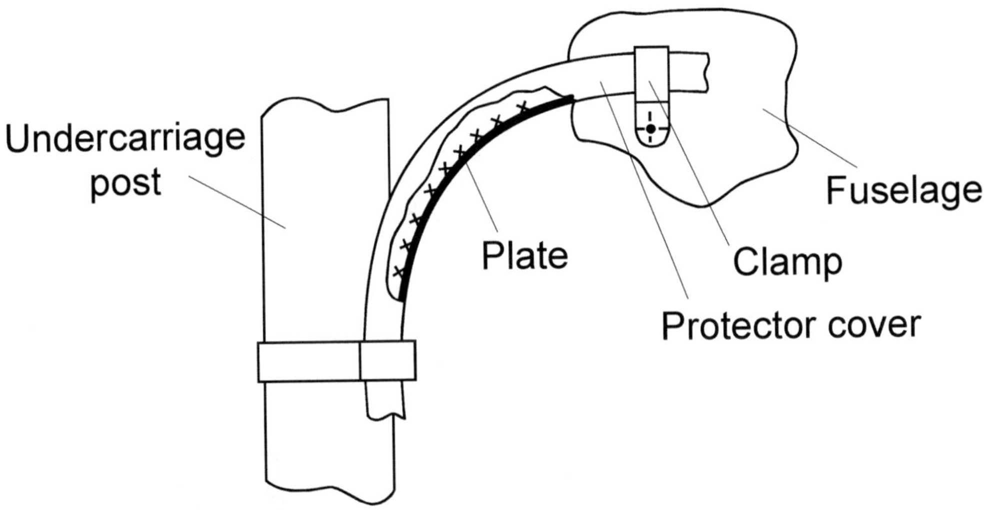

Fig. 4.2. The change of plasticizer concentration in mass % in cable isolation dismounted from articles after different periods of exploitation.

Protections from mechanical, thermal and other damages is a must for cables which are:

1. Connected to folding-back panels, desks, etc.;
2. Designed for mounting in zones with high temperatures exceeding the one permitted by cable insulation heat resistance;
3. Mounted near aggregates and pipelines of oil, hydraulic, fuel systems, and in places subjected to gas influence;
4. Mounted near sharp edges, rudders and other places affected by external mechanical tensions.

Moreover gaps must be provided between braids and construction elements.
Defects cleared out in exploitation of BEN cables are the following:

- Soiling of cables and protective coverings of braids and cables;
- Changing of insulation colours (become darker);
- Mechanical damages of cables and braids.

4.2 Test Results of Board Cable Inspection with PVC-plasticate Insulation Striped from Articles

The following tests of board cables with PVC-plasticate insulation, dismounted from articles, were performed:

- Measurements of insulation mass after heating at 350°C during 25 min, % (see Tables 4.2 and 4.4)

$$\frac{m_0 - m_{350^\circ}}{m_0} = \Delta G;$$

- Measurements of cable diameter (see Table 4.3);
- Determination of electrical resistance of cable insulation at 70% relative humidity in air and in water (see Table 4.5);
- Visual inspection in order to find mechanical damages (breaks), colour change, soilings, etc. (see Table 4.5).

Tests were carried on cable fragments of 30 ÷ 50 cm long. Cable fragments were labeled by a label with the cable or braid brand and dislocation in the article signed. Values of ΔG parameter were determined as a mean-arithmetic of three measurements, with pointing out maximal and minimal values of ΔG in some cases. Values of ΔG were measured at the beginning and in the end of cable for most of cables. In some braids (CP28, CL28, C158, DP30) these values are sufficiently different, which is bound to different temperature regimes of exploitation.

Table 4.2. Test results of measurements of ΔG parameter for BWPV cables dismounted from articles 'I' and 'II'.

No.	Braid Label	Cable label	Values of ΔG parameter, %				
			Average	Minimum	Maximum	At the beginning of the cable	At the end of the cable
1	2	3	4	5	6	7	8
Article 'II', the leader							
1	C156	11M6	—	65.96	68.57	68.4	67.0
2	C156	11M6	—	—	—	68.6	67.9
3	C156	11M6	—	—	—	68.2	66.2
4	C156	11M6	—	—	—	66.0	67.0
5	Cable 2	—	—	63.03	65.68	—	—
6	Cable 2	—	—	—	—	63.0	63.7
7	Cable 2	—	—	—	—	63.5	63.4
8	Cable 2	—	—	—	—	—	—
9	Cable 2	—	—	—	—	64.8	—
10	Cable 2	—	—	—	—	63.6	64.1
11	Cable 2	—	—	—	—	—	64.0
12	Cable 2	—	—	—	—	—	—
13	Cable 2	—	—	—	—	—	—
14	Cable 2	—	—	—	—	—	—
15	Cable 2	—	—	—	—	—	—
16	Cable 2	—	—	—	—	64.8	—
17	CP28	2IS28	—	59.99	65.68	60.0	—
18	CP28	IS28	—	—	—	—	63.6
19	CP28	1US28	—	—	—	60.3	65.9
20	CP28	1US29	—	—	—	60.1	—
21	CP28	1US27	—	—	—	61.9	—
22	CP28	1SS8	—	—	—	60.3	—
23	CP28	1SS6	—	—	—	60.1	—
24	CP28	b/b	—	—	—	63.1	—
25	CP28	b/b	—	—	—	64.1	64.4
26	C472	2US13	—	64.95	68.51	65.4	65.0
27	C472	2US10	—	—	—	68.4	68.5
28	P145	—	—	—	—	—	—
29	P145	—	—	71.30	72.32	72.3	—
30	P144	—	—	70.77	72.17	—	—
31	P144	—	—	—	—	71.5	72.2
32	P144	—	—	—	—	71.0	71.3
33	P144	—	—	—	—	—	—
34	P144	—	—	—	—	70.1	—
35	P162	—	—	64.98	70.03	67.2	—
36	P162	—	—	—	—	—	68.0
37	P162	—	—	—	—	67.8	68.2
38	P162	—	—	—	—	66.7	—
39	P162	—	—	—	—	—	—

40	P162	—	—	—	—	68.0	—
41	P162	—	—	—	—	—	—
42	P162	—	—	—	—	—	—
43	P162	—	—	—	—	63.2	70.0
44	P162	—	—	—	—	—	65.0
45	P162	—	—	—	—	67.4	—
46	P162	—	—	—	—	—	—
47	L53	—	—	—	—	—	—
48	L53	—	—	62.82	67.52	64.9	—
49	L53	—	—	—	—	—	—
50	L53	—	—	—	—	64.8	—
51	L53	—	—	—	—	—	65.5
52	L53	—	—	—	—	65.6	64.5
53	L53	—	—	—	—	—	63.5
54	L53	—	—	—	—	—	67.2
55	L53	—	—	—	—	—	67.5
56	L53	—	—	—	—	62.8	—
57	L53	—	—	—	—	—	—
58	L53	—	—	—	—	—	—
59	L53	—	—	—	—	66.8	—
60	L53	—	—	—	—	—	—
61	L53	—	—	—	—	—	—
62	L53	—	—	—	—	63.4	—
63	L53	—	—	—	—	—	—
64	L53	—	—	—	—	—	—
65	L53	—	—	—	—	—	—
66	L53	—	—	—	—	—	—
67	L53	—	—	—	—	—	—
68	L53	—	—	—	—	—	—
69	L53	—	—	—	—	—	64.1
70	L53	—	—	—	—	65.1	—
71	C276	4MW6	—	64.55	69.58	66.7	68.2
72	C276	4MW5	—	—	—	64.6	69.6
73	C276	4MW4	—	—	—	—	—
74	CL28	1SW6	—	61.23	70.4	65.7	70.4
75	CL28	1SW6	—	61.83	—	62.0	61.2
76	CL28	b/b	—	61.17	—	62.6	61.8
77	CL28	IUW20	—	61.47	—	61.2	62.6
78	CL28	2UW21	—	6221	—	62.2	63.1
79	CL28	1YW21	—	61.48	—	61.5	63.1
80	CL28	1UW20	—	64.34	—	61.9	61.3
81	CL28	2SW5	—	61.01	—	61.0	61.6
82	CL28	1UW17	—	61.39	—	61.2	62.6
83	Cable 15	—	—	61.39	64.5	63.0	—
84	Cable 15	—	—	—	—	61.4	64.3
85	Cable 15	—	—	—	—	—	63.7
86	Cable 15	—	—	—	—	64.6	63.4

87	Cable 15	—	—	—	—	—	—	—
88	Cable 15	—	—	—	—	—	61.1	62.1
89	Cable 15	—	—	—	—	—	62.4	—
90	10-3Pr	—	—	63.91	66.98	66.6	—	
91	10-3Pr	—	—	—	—	67.0	66.3	
92	10-3Pr	—	—	—	—	65.9	67.0	
93	10-3Pr	—	—	—	—	64.5	63.9	
94	10-3Pr	—	—	—	—	64.2	—	
95	C218	3MU2	—	71.93	73.30	71.9	—	
96	C218	3MU4	—	—	—	72.0	73.3	
97	C218	37-G4	—	—	—	72.9	72.6	
98	C218	b/b	—	—	—	73.1	—	
99	C158	—	—	60.62	71.62	67.1	68.6	
100	C158	—	—	—	—	71.6	67.9	
101	C158	—	—	60.92	—	60.6	66.4	
102	C158	—	—	61.97	—	62.0	65.1	
103	C158	—	—	63.12	—	66.5	—	
104	DP130	3MP6	—	60.05	64.01	62.0	65.1	
105	DP130	3MP11	—	61.30	—	63.1	61.3	
106	DP130	3MP11	—	—	—	—	—	
107	DP130	3MP12	—	—	—	64.0	63.2	
108	DP130	3MP13	—	61.99	—	63.7	62.0	
109	S108	—	—	64.94	68.13	65.6	—	
110	S108	—	—	—	—	65.7	64.9	
111	S108	—	—	—	—	68.1	67.3	
112	S108	—	—	—	—	67.3	—	
113	S108	—	—	—	—	66.4	66.5	
114	S108	C450	—	67.57	67.66	67.7	67.6	
115	S108	C360	—	69.57	69.56	70.0	69.6	
116	S108	C361	—	68.49	69.52	68.5	69.5	
117	S108	UW10	—	65.79	67.90	67.9	65.8	
118	S108	UW11	—	65.74	67.56	67.6	65.7	
119	R145	1RU2	—	62.25	66.83	—	—	
120	R145	1RU3	—	—	—	68.6	66.5	
121	R145	3RU1	—	—	—	—	—	
122	R145	3RU2	—	—	—	66.8	64.1	
123	R145	3RU16	—	—	—	65.3	63.8	
124	R145	1RU5	—	—	—	64.3	62.2	
125	R145	b/b	—	—	—	65.1	66.1	
126	R145	b/b	—	—	—	—	—	
Article 'I', No.701								
127	F-2, AP-15SP	AP308	67.2	—	—	—	—	
128	Same	AP320	69.7	—	—	—	—	
129	Same	AP307	68.7	—	—	—	—	
130	Same	AP315	65.9	—	—	—	—	
131	Same	AP321	64.9	—	—	—	—	

132	Same	AP322	63.2	—	—	—	—
133	Same	AP314	65.3	—	—	—	—
134	Same	AP340	67.8	—	—	—	—
135	Same	AP430	66.3	—	—	—	—
136	Same	AP313	63.0	—	—	—	—
137	Same	AP429	66.4	—	—	—	—
138	Same	AP305	63.0	—	—	—	—
139	Same	AP302	63.8	—	—	—	—
140	Same	AP312	67.6	—	—	—	—
141	Same	AP311	66.3	—	—	—	—
142	Same	AP301	65.4	—	—	—	—
143	Same	AP310	64.8	—	—	—	—
144	Same	AP339	64.2	—	—	—	—
145	Same	AP318	65.1	—	—	—	—
146	Same	AP319	64.8	—	—	—	—
147	Same	AP317	64.2	—	—	—	—
148	Same	AP304	65.2	—	—	—	—
149	Same	AP303	63.1	—	—	—	—
150	Same	AP316	64.7	—	—	—	—
151	F-2, AP-15USP	AN430	69.4	—	—	—	—
152	Same	AP302	63.9	—	—	—	—
153	Same	AP304	68.7	—	—	—	—
154	Same	AP307	67.0	—	—	—	—
155	Same	AP305	68.6	—	—	—	—
156	Same	AP340	71.8	—	—	—	—
157	Same	AP213	65.9	—	—	—	—
158	Same	AP310	72.1	—	—	—	—
159	Same	AP318	65.3	—	—	—	—
160	Same	AP319	65.5	—	—	—	—
161	Same	AP303	64.5	—	—	—	—
162	Same	AP322	66.9	—	—	—	—
163	Same	AP321	64.0	—	—	—	—
164	Same	AP317	70.4	—	—	—	—
165	Same	AP312	71.4	—	—	—	—
166	Same	AP429	66.2	—	—	—	—
167	Same	AP339	66.7	—	—	—	—
168	Same	AP311	66.5	—	—	—	—
169	Same	AP301	65.9	—	72.9	—	—
170	Same	AP320	67.6	—	—	—	—
171	Same	AP315	66.6	—	—	—	—
172	Same	AP314	66.5	—	—	—	—
173	Same	b/b	62.8	—	—	—	—
174	P.S., C474	1US8	66.6	—	—	—	—
175	Z.S., C267	2US5	68.8	—	—	—	—

176	Same	1US22	67.8	—	—	—	—
177	Same	1US15	67.3	—	—	—	—
178	Same	1US19	67.2	—	—	—	—
179	Same	1US5	67.6	—	—	—	—
180	Same	1US15	65.7	—	—	—	—
181	Same	1US21	68.4	—	—	—	—
182	Z.S., C277	4US3	67.3	—	—	—	—
183	Same	3US3	68.3	—	—	—	—
184	Same	TK	66.3	—	—	—	—
185	Z.S., C247	4MS2	68.8	—	—	—	—
186	Same	4MS3	67.6	—	—	—	—
187	Same	3MS	68.9	—	—	—	—
188	Z.S., C248	3MS2	67.3	—	—	—	—
189	Same	3MS3	68.2	—	—	—	—
190	Same	MS	68.5	—	—	—	—
191	C-2L, 10-3	b/b	68.0	—	—	—	—
192	Same	b/b	68.0	—	—	—	—
193	C-2P, 10-3	b/b	68.8	—	—	—	—
194	Same	b/b	69.0	—	—	—	—
195	G.L., C2	2US9	65.6	—	—	—	—
196	Same	3US6	68.7	—	—	—	—
197	G.L., C3	1US9	64.5	—	—	—	—
198	Same	3US9	65.1	—	—	—	—
199	4 double DL66	EG5	65.2	—	—	—	—
Article 'I', No. 604							
200	F-2, AP-15	AP306	66.0	—	—	—	—
201	Same	AP320	65.9	—	—	—	—
202	Same	AP307	—	—	—	—	—
203	Same	AP315	—	—	—	—	—
204	Same	AP321	65.7	—	—	—	—
205	Same	AP322	64.8	—	—	—	—
206	Same	AP314	—	—	—	—	—
207	Same	AP340	65.7	—	—	—	—
208	Same	AP430	63.6	—	—	—	—
209	Same	AP313	65.5	—	—	—	—
210	Same	AP429	64.6	—	—	—	—
211	Same	AP305	65.3	—	—	—	—
212	Same	AP302	65.9	—	—	—	—
213	Same	AP312	65.5	—	—	—	—
214	Same	AP311	63.9	—	—	—	—
215	Same	AP301	64.7	—	—	—	—

216	Same	AP310	64.6	—	—	—	—
217	Same	AP339	62.8	—	—	—	—
218	Same	AP318	64.3	—	—	—	—
219	Same	AP319	64.3	—	—	—	—
220	Same	AP317	63.4	—	—	—	—
221	Same	AP304	64.9	—	—	—	—
222	Same	AP303	64.4	—	—	—	—
223	Same	AP316	—	—	—	—	—
224	AP15 USP	AP430	64.1	—	—	—	—
225	Same	AP302	—	—	—	—	—
226	Same	AP304	64.5	—	—	—	—
227	Same	AP307	62.6	—	—	—	—
228	Same	AP305	67.8	—	—	—	—
229	Same	AP340	67.1	—	—	—	—
230	Same	AP313	65.6	—	—	—	—
231	Same	AP310	67.2	—	—	—	—
232	Same	AP318	64.7	—	—	—	—
233	Same	AP319	66.4	—	—	—	—
234	Same	AP303	66.3	—	—	—	—
235	Same	AP322	65.9	—	—	—	—
236	Same	AP321	66.7	—	—	—	—
237	Same	AP317	64.7	—	—	—	—
238	Same	AP312	64.1	—	—	—	—
239	Same	AP429	66.0	—	—	—	—
240	Same	AP339	64.0	—	—	—	—
241	Same	AP311	65.5	—	—	—	—
242	Same	AP301	66.3	—	—	—	—
243	Same	AP320	66.0	—	—	—	—
244	Same	AP315	—	—	—	—	—
245	Same	AP314	—	—	—	—	—
246	Same	b/b	68.3	—	—	—	—
247	C474	1US8	—	—	—	—	—
248	C267	2US5	—	—	—	—	—
249	C248	3MS2	68.2	—	—	—	—
250	Same	3MS3	68.7	—	—	—	—
251	Same	MS	67.0	—	—	—	—
252	DL-66	EG-5	67.1	—	—	—	—

Table 4.3. Diameter of the cable No. 24 of CP28 braid.

No.	Cable diameter, mm	
	At the cable beginning (ΔG = 60.3%)	**At the cable end (ΔG = 65.7%)**
1	2.03	2.15
2	2.01	2.14
3	2.04	2.11
4	2.05	2.14
5	2.03	2.10
6	2.07	2.13
7	2.05	2.12
8	2.04	2.15
9	2.01	2.17
10	2.05	2.14
11	2.03	2.15
12	2.04	2.17
13	2.07	2.11
14	2.05	2.10
15	2.04	2.15
16	2.06	2.17
17	2.01	2.16
18	2.03	2.12
19	2.05	2.14
20	2.03	2.15
21	2.04	2.17
22	2.05	2.14
23	2.04	2.15
24	2.03	2.17
25	2.02	2.14
Average	2.04	2.14

Fig. 4.3 shows the change of relative plasticizer mass in insulation of cables, dismounted from articles after different periods of exploitation.

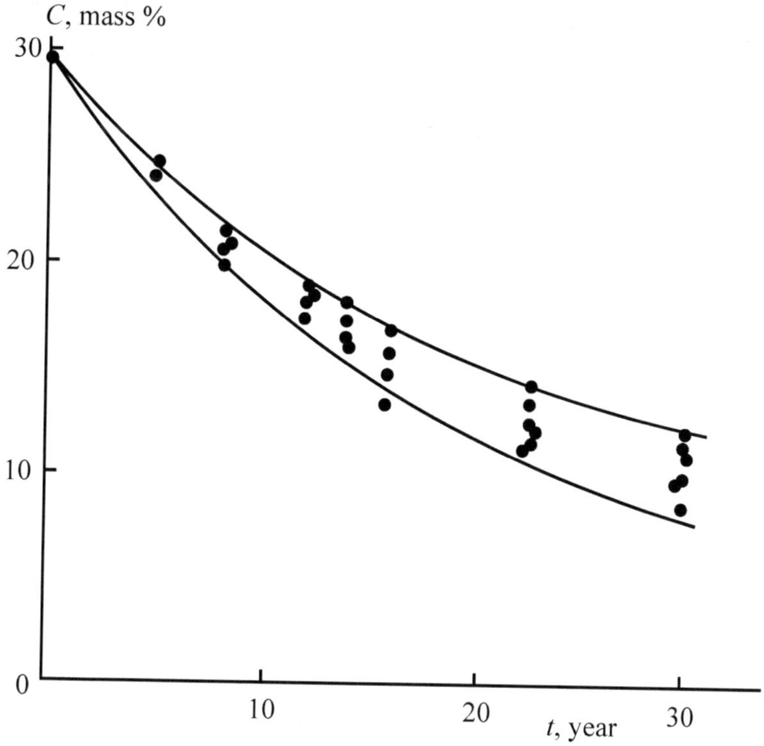

Fig. 4.3. Kinetic curves of the change of mass and volume of pores in PVC-plasticate isolation depending on exploitation time.

Plasticizer mass in insulation samples was measured by following techniques:

- Thermogravimetrically by heating insulation sample at 350°C during 25 min. Plasticizer concentration in the material was calculated by the following equation:

$$C_{pl} = \Delta G - 0.58C_{PVC} - \beta C_{ad}, \tag{40}$$

where C_{PVC} and C_{ad} are concentrations of PVC and industrial additives in the initial material, respectively, in mass parts; β is the rate of additive thermodecomposition in the present regime;

- By spectrum technique, by both UV-spectra, preliminarily solving insulation sample in tetrahydrofuran and calculating plasticizer concentration by the equation (34), and IR-spectra of reflection by absorption band intensity at 1720 cm^{-1}.

Plasticizer mass loss from insulation materials during exploitation during 30 years in the middle region (of Russia) reaches 70%, and the relative mistake of plasticizer mass calculation increases with exploitation time in this case.

Cable diameter does also decrease during exploitation because of shrinkage, however the initial diameter of cables has not been known in all cases, that is why these results are not shown.

5 PROGNOSIS

5.1 Development of Methods of Mathematical Description and Determination of PVC-plasticate Insulation Break-down

The break-down state is determined by placing a 1 meter fragment of cable into water and applying electric current of 10 V to cable and to volume. Shorting testifies the insulation break-down provided by existence of through holes (channels) in it, filled with water. Such measurements, which acquire a great amount of wires, cannot be used for rapid estimation of board electric network capability to work during exploitation on flying facilities.

In this connection the following tasks were set:

- To develop a method of mathematical description of the break-down state, i.e. to characterize it quantitatively by parameters of physical and chemical properties, which may be determined experimentally;
- To develop methods of experimental determination of the break-down state using microamounts of cable insulation, i.e. to create an express method for determination of the break-down state;
- To develop a method for prognosis of remaining resource of cables with PVC-plasticate insulation based on the stated mechanism of ageing.

5.1.1 Methods of Mathematical Description of PVC-plasticate Insulation Break-down

Plasticized PVC represents the molecular network, disposed in a plasticizer. Crystallites or globules are lattice points. The predominant process of PVC-plasticate ageing under exploitation conditions (< 40°C), which causes the change of exploitation properties, is diffusion desorption of plasticizer accompanied by shrinkage of the isolating covering. However, PVC-plasticate inclination to the volume decrease is limited by a kinetic factor and isometry of the article. This provides formation of pores and accumulation of elastic deformation energy in the material. Pore volume will grow in the material during ageing. Connecting with each other, pores can form a system of through channels (flow channels), which connect the surface and internal border of the insulation covering and cause covering break-down.

The break-down state may be explained from positions of the flow theory, which sets the existence of a critical part (flow threshold) of pores, at which pores form connected system along the whole volume of the covering.

Let us consider strict description of the model of PVC-plasticate and formulate some relative mathematical results.

1. At the initial moment Poisson distribution (with intensity μ_0) of O_i centers of 'congenital' pores is set in the volume V with R^3. Pores S_i connected with O_i centers are

independent equally distributed open [sets]. Initial pore set $D_v = \bigcup_i S_i$ is divided into linearily connected components, which should be named as 'congenital' pores.

Volumetric concentration of the pore set at time $t = 0$ is the following:

$$\lambda_0 = \lim_{V \to R^3} \frac{Vol(D_v)}{Vol(V)}. \tag{41}$$

The limit can be easily calculated in terms μ_0 and distribution S_i. For example, if $S_i = \{X: |X - O_i| \le \xi_i\}$ and $P\{\xi_i E(r_1 r + dr)\} = P(r)dr$,

$$\lambda_0 = \exp\left\{-\mu_0 \int_0^{\infty} 4\pi r^2 P(r)dr\right\}. \tag{42}$$

2. Pore development during ageing is more suitable for characterization by the following function:

$$\lambda(t) = \lim \frac{Vol(D_v(t))}{Vol(V)}, \tag{43}$$

where $D_v(t)$ is the set of pores at time t. Time dependence of the distribution S_i should be taken into account also.

The most simple ageing model is the following. Micropores appear in material with the rate v_0, and the more so all pores existing in the volume at time t grow (by normal to $\partial D_j(t)$) with the rate v_1. Initial pores S_i (0) are balls possessing radius distribution density $P(r)$ and intensity μ_0 of center appearance.

Diffusion desorption proceeds at constant temperature T, and rates v_0 and v_1 depend on T according to Arrhenius law:

$$v_0(T) = C_0 \exp\left(-\frac{E_0}{T}\right), \tag{44}$$

$$v_1(T) = C_1 \exp\left(-\frac{E_1}{T}\right). \tag{45}$$

It can be shown that

$$\lambda(t) = \exp\{-P_4(t)\}, \tag{46}$$

here $P_4(t)$ is the order four polynomial, which coefficients are clearly expressed by μ_0, ν_0, ν_1 and $P(r)$.

If $\nu_1 >> \nu_0$ and t is not too long, the main role is played by linear terms of the polymonial, which depend on μ_0, ν_1 and $P(r)$.

3. It does not matter what mechanism of pore development is on the run, volumetric concentration of pores $\lambda(t)$ grows in any case. Using combinatorial methods, contour technique in particular, we can find that at some critical value λ* flow channels occur. The correspondent moment t*; $\lambda\left(t^*\right) = \lambda^*$ is the flow threshold reached.

Mathematical modelling shows that coordinational number of pores does not depend on the form and exists in the range of 3.0 ± 0.2, however volumetric concentration of pores may change in a wide range, when the flow threshold is reached: from 0.28 for balls of equal radii to > 0.01 for needle-shaped pores (longily elongated rotation ellipsoids).

4. The result of the point 3 means that the moment of the polymer covering break-down must be close to t* and corresponding function must be the indicating one:

$$P_\xi(x) = 0 \ \text{for} \ x < t^* \ \text{and} \ P_\xi(x) = 1 \ \text{for} \ x > t^*.$$

Deviation from this law may be bound to the existence of 'congenital' pores.

5. When the critical concentration λ* is reached, mechanical and physical properties of PVC-pasticate, as well as its geometrical properties (meaning the flow theory), are changed.

In such a manner the amount and total volume of pores in PVC-plasticate increase at ageing. When the critical pore volume is reached, the system of connected flow channels occurs, forming the break-down state. The value of critical pore volume can be theoretically calculated taking into account pores formed and the amount of initial 'congenital' pores.

Taking into account that insulating covering contacts metal wire of a cable, shrinkage at plasticizer desorption will proceed in radial direction only. In accord to geometrical fundamentals we may accept that pores formed will possess the form of rotation ellipsoids stretched along the interface of polymer contact with metal. The break-down state displays ratio of rotation ellipsoid axises as 2:1. Such pores display critical value of volumetric concentration of ~5%.

5.1.2 Development of Methods for Experimental Break-down Determination

Analysis of the experimental data showed that two parameters can be recorded with a satisfactory occuracy at PVC-plasticate ageing: remaining amount of plasticizer in the material and total volume of pores and their distribution by diameters. As it is seen in the part 4.1, both these parameters may be applied to quantitative characterization of the break-down state.

Two following methods are recommended for determination of the remaining amount of plasticizer in insulation samples dismounted from articles:

- ***Thermogravimetric technique:*** is used at sample heating at 350°C during 25 min. Calculation of plasticizer amount is performed by equation (40). Preliminary leveling is required. Average duration of analysis is 30 min, relative mistake is ± 4%, minumum of the material amount is 50 mg.
- ***IR-spectrum technique:*** is performed by reflection spectra by intensity of absorption band at 1720 cm^{-1} frequency. Preliminary leveling is required. Average duration of analysis is 5 min, relative mistake is ± 5%. Minimum of the material square is 20 μm^2. Sample is not destroyed during the analysis.

Fig. 5.1 shows curves of mass change and total pore volume in PVC-plasticate insulation depending on exploitation time under natural conditions. The latter value was calculated as difference between mass loss and insulation volume decrease at shrinkage:

$$V_{pore} = \frac{\Delta m_{pl}}{\rho_{pl}} - \Delta V, \tag{47}$$

here V_{pore} is pore volume; Δm_{pl} and ρ_{pl} are the mass of desorbed plasticizer and plasticizer density, respectively; ΔV [1] is the change of article volume.

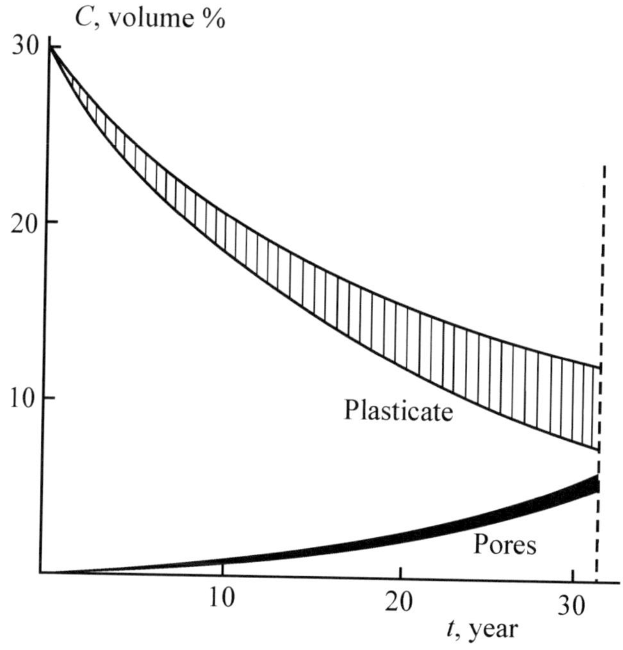

Fig. 5.1

[1] The value of ΔV was determined for a part of insulation samples only dismounted from articles.

This Fig. 5.1 also shows the straight line (broken line), correspondent to 10% of break-downs obtained in cables dismounted from articles. To this line 8-11% of remaining plasticizer mass and 5-6% of pore volume are corresponded.

In such a manner we may state that critical value of volumetric pore concentration (λ^*), to which the break-down state relates, is 5.5 ± 0.5%.

Total volume of pores in the break-down state may be directly determined in experiment by mercury porometry.

As it is seen in the Table 3.3, total pore volume in the break-down state is 6.5%, i.e. it is practically equal to the value obtained from determinations of plasticizer mass losses with calculation of sample shrinkage. Thus, experimentally determined critical value of the pore concentration is the same as the value, calculated theoretically with taking into account the form of pores formed and the amount of initial 'congenital' pores.

5.1.3 Prognosis of the Residual Wire Resource with PVC-plasticate Insulation

In the previous parts we determined that the break-down state displays 8-11% of remaining plasticizer mass and 5-6% of the pore volume formed. Basing on these data we can estimate the remaining resource time of cable according to the Fig. 5.1. If the mass of plasticizer in the material is known, we can calculate the time of exploitation, correspondent to accepted equivalent temperature (15-20°C). If calculated exploitation time equals to the real one, the time of remaining resource is determined as the difference of t^* (the break-down state time) and the time of previous exploitation. If calculated exploitation time differs from the real one by more than 5%, it is necessary to determine equivalent temperature (T_e) once again by the following equation, deduced from the equation (35):

$$T_e = -\frac{E}{R}\left\{\ln\left[\frac{1}{\tau_0^{0.62}}\sum_{j=1}^{n}\Delta\tau_j^{0.62}\exp\left(-\frac{E}{RT_j}\right)\right]\right\}^{-1}, \tag{48}$$

where E is activation energy of diffusional desorption of plasticizer; τ_0 is mean-statistic time (in hours) of the present exploitation duration; $\Delta\tau_j$ is duration of temperature range with average temperature T_j.

The temperature T_e determined must be applied in further calculations.

Macropores occur, when the break-down state is reached. Their sizes are comparable with covering thickness, that is why testing of the insulation on the content of macropores is a must, when the remaining resource, close to the break-down, is determined.

6. Conclusion

1. Basing on the data from literature and experimental data it was shown that predominant process in ageing of insulation made from PVC-plasticate at < 40°C that causes the break-down is diffusional desorption of plasticizer. If exploitation tempratures are > 100°C (in engine compartments and pipes of anti-icing system), beside diffusional desorption of plasticizer we obtain dehydrochlorination and thermoxidative degradation.

2. Methods of determination of the remaining plasticizer amount in PVC-plasticate were developed: thermogravimetric technique concluded in determination of the material mass after heating at 350°C during 25 min; spectral techniques according to absorption bands at 230 and 274 nm in PVC-plasticate solution in tetrahydrofuran (UV-spectrometer) and by 1720 cm^{-1} absorption band in spectra obtained on IR-spectrometer.
3. It was shown that diffusional desorption of plasticizer at ageing under natural conditions and in rapid tests (>100°C) is accompanied by decrease of insulation volume (shrinkage), that does not completely compensate the insulation mass loss and causes pore formation in the material. Mercury porometry technique made possible the determination of the total pore volume and pore distribution by diameters in PVC-plasticate samples after different periods of ageing. It was shown that pore volume and their sizes increase during ageing, the latter from 100 to 500 nm. Pores with 3 μm sizes were observed at the break-down state, what is the characteristic feature of the break-down.
4. The method of mathematical description of the break-down in PVC-plasticate insulation was developed on the basis of the flow theory. It was shown that the break-down state is characterized by the critical pore volume, at which formation of the connected flow channel system occurs. These channels provide phase water transfer. The value of the critical pore volume was calculated with taking into account the form of pores formed (rotation ellipsoids with 2:1 ratio of axes), and the amount of initial pores, equal 5.5 ± 0.5%, was also determined.
5. Methods were developed for the prognosis of the remaining resource of cables covered by PVC-plasticate insulation. These methods were developed with the help of deduced kinetic equation describing the mass change of plasticizer with the ageing time and equivalent exploitation temperature.
6. Basing on the data obtained it was suggested to revise the interrepair times of flying facilities and exploitation times of cables with PVC-plasticate insulation. It is necessary to take into account the initial concentration of pores in the material, when exploitation times of cables with PVC-plasticate insulation are determined.

References

1. GOST 21126-75, *Rapid Test Methods on Durability and Retention in Aggressive Media. General Positions*, in Russian.
2. GOST 9.707-81, *Polymer materials. Rapid Test Methods of Climatic Ageing,* in Russian.
3. K.S. Minsker, S.V. Kolesov and G.E. Zaikov. *Degradation and Stabilization of Vinylchloride Based Polymers*, Pergamon Press, Oxford-New York-Toronto, p. 78-82 (1988).
4. K.S. Minsker, S.V. Kolesov and G.E. Zaikov. *Degradation and Stabilization of Vinylchloride Based Polymers*, Pergamon Press, Oxford-New York-Toronto, p. 187-203 (1988).
5. K.S. Minsker, M.T. Abdulin, R.R. Gizatulin and A.L. Buchachenko. *Doklady Acad. Nauk SSSR*, **276**, 1181-1184 (1984), in Russian.
6. K.S. Minsker and G.E. Zaikov in: *Polymer Yearbook*, **4**, pp. 222-232, Harwood Academic Publishers (1987).
7. M.A. Zhurvlev and V.B. Ivanov. *Eur. Polym. J.*, **25**, 391 (1989).

8. K.S. Minsker, S.V. Kolesov and G.E. Zaikov. *Degradation and Stabilization of Vinylchloride Based Polymers*, Pergamon Press, Oxford-New York-Toronto, p. 112-121 (1988).
9. F. Tudos, B. Ivan, T. Kelen and I.P. Kennedy. *Dev. Polym. Degrad.*, **6**, N. Grassier (Ed.), Elsevier, London, pp. 147-158 (1985).
10. A.A. Caraculacu. *Pure Appl. Chem.*, **53**, 385 (1981).
11. W.H. Starnes. *Dev. Polym. Degrad.*, **3**, 135 (1981).
12. D. Braun. *Pure Appl. Chem*., **53**, 549 (1981).
13. D. Braun. *Dev. Polym. Degrad.*, **3**, 101 (1981).
14. B. Ivan in: *Polymer Durability: Degradation, Stabilization and Lifetime Prediction*, R. Clough, N.C. Billingham and K.T. Gillen (Eds.), pp. 19-32, *Adv. Chem. Ser.*, American Chemical Society, Washington, DC (1996).
15. D. Braun, B. Bohringer, B. Ivan, E. Mori, T. Kelen and F. Tudos. *Angew. Makromol. Chem.*, **143**, 165 (1986).
16. K.S. Minsker, D.V. Kozatchenko, R.G. Abdullina, R.B. Kovler and Al.Al. Berlin. *Highmolec. Comp.*, **A15**, 866 (1973), in Russian.
17. I. Boissel. *J. Appl. Polym. Sci.*, **21**, 855, (1977).
18. A. Michel, E. Gastaneda and A. Guyot. *J. Macromol. Sci.*, **A12**, 227 (1978).
19. R.F. Tataulin. *Kinetics and Catalysis*, **19**, 228 (1978).
20. A. Caraculacu and E. Berdadea (Eds.). *J. Polym. Sci.: Polym. Chem.*, **15**, 611 (1977).
21. K.S. Minsker, V.V. Liscitzky and N.V. Davivenko. *Plast. Massy*, U, 16 (1982), in Russian.
22. K.S. Minsker, Al.Al. Berlin and V.V. Liscitzky. *Highmolec. Comp.*, **B18**, 54 (1976).
23. K.B. Abbas and E.M. Sorvik. *J. Appl. Sci.*, **20**, 2395 (1976).
24. K.S. Minsker, Al.Al. Berlin, V.V. Liscitzky and S.V. Kolesov. *Highmolec. Comp.*, **A19**, 32 (1977).
25. D. Braun, A. Michel and D. Sonderhof. *Europ. Polymer J.*, **17**, 49, (1981).
26. F.A. Bovey, F.C. Shilling and W.H. Starkes. *Amer. Chem. Sci.*, Div. Polymer Prepr., **20**, 160 (1979).
27. S. Grawlly and I.C. McNeill. *J. Polym. Sci., Polym. Chem. Ed.*, U, 2593 (1978).
28. I. Millan, E. Madruga and G. Martinez. *Angew. Makromol. Chemie*, **45**, 177 (1975).
29. G. Martinez and I. Millan. *Angew. Makromol. Chemie*, **75**, 215 (1979).
30. G. Martinez. *Macromol. Chem.*, **180**, 2937 (1979).
31. L. Milan, G. Martinez and C. Mijangos. *J. Polymer Sci., Polym. Chem. Ed.*, **18**, 505 (1980).
32. L. Milan, G. Martinez and C. Mijangos. *Polymer Bull.*, **5**, 407 (1981).
33. K.S. Minsker, V.V. Liscitzky and G.E. Zaikov. *Highmolec. Comp.*, **A23**, 483 (1981), in Russian.
34. V.V. Liscitzky, S.V. Kolesov, R.F. Gataullin and K.S. Minsker. *J. Anal. Chem.*, **33**, 2202 (1978).
35. S.V. Kolesov and K.S. Minsker. *Highmolec. Comp.*, **A25**, 1587 (1983), in Russian.
36. I. Bauer and A. Sabel. *Angew. Makromol. Chemie*, **47**, 15 (1975).
37. D. Braun and M. Wolf. *Angew. Makromol. Chemie*, **70**, 71 (1978).
38. D. Braun and D. Sonderhof. *Europ. Polym. J.*, **18**, 141 (1982).

39. K.S. Minsker, S.V. Kolesov and G.E. Zaikov. *Degradation and Stabilization of Vinylchloride Based Polymers*, Pergamon Press, Oxford-New York-Toronto, p. 104 (1988).
40. V.I. Zegelman, V.V. Liscitzky, V.A. Titova, S.I. Miroshnichenko, A.P. Alexandrov, V.I. Genkin and V.A. Popov. *Doklady Acad. Nauk SSSR*, **277**, 896 (1984), in Russian.
41. V.I. Zegelman, V.A. Titova, V.C. Etlis, V.A. Popov and E.G. Pomerantcheva. *Doklady Acad. Nauk SSSR*, **278**, 399 (1984), in Russian.
42. P. Kohn, C. Marechal and I. Verdes. *Anal. Chem.*, **51**, 1000 (1979).
43. W.I. Bengough and P.J. Chamy. *Thesis IUPAC Sympos.*, p. 284, Amherst, USA (1982).
44. A. Guyot and M. Bert. *Pure Appl. Chem.*, **53**, 401 (1981).
45. B. Ivan, T. Kelen and F. Tudos. *Vinyl: A Material for the Future*, App., pp. 1-12, SPE Meeting, Montreal, September 15-16 (1988).
46. Z. Mayer, B. Obereigner and D. Zim. *J. Polym. Sci.*, **12**, 239 (1974).
47. B.B. Troitsky and L.S. Troitskaya. *Highmolec. Comp.*, **A20**, 1443 (1978), in Russian.
48. K.S. Minsker and G.G. Fedoseyeva. *Degradation and Stabilization of Polyvinylchloride*, Khimia, Moscow, 272 p. (1979), in Russian.
49. B. Ivan, T. Kelen and F. Tudos in: *Degradation and Stabilization of Polymers*, H.H.G. Jellinek (Ed.), **2**, pp. 483-714, Elsevier Sci. Publ. Co. (1989).
50. I. Klaric, U. Rije and T. Kovacic. *Chem. Ind.*, **36**, 145 (1982).
51. S.G. Entelis and R.P. Tiger. *Reaction Kinetics in the Liquid Phase*, Khimia, Moscow (1973), in Russian.
52. I. Bauer and A. Sabel. *Angew. Makromol. Chemie*, **47**, 15 (1975).
53. T. Kelen, B. Ivan and F. Tudos in: *Advances in the Stabilization and Controlled Degradation of Polymers*, A.V. Patsis (Ed.), **1**, pp. 39-51, Technomic, Lancaster (1989).
54. K.B. Abbas. *J. Macromol. Sci.*, **A12**, 479 (1978).
55. P. Saiks. *Mechanisms of Reactions in Organic Chemistry*, Khimia, Moscow (1977), in Russian.
56. W.H. Stranes, Jr. and S. Girois. *Polymer Yearbook*, **12**, 105 (1997).
57. W.H. Stranes. *Amer. Chem. Soc., Polym. Prepr.*, **21**, 138 (1980).
58. Yu.V. Moiseev and G.E. Zaikov. *Chemical Stability of Polymers in Aggressive Media*, pp. 204-215, Khimia, Moscow (1979).
59. R.A. Parko and V.S. Pudov. *Highmolec. Comp.*, **A18**, 864 (1976).
60. S.B. Pausch, C.A. McKalen and A.R. Bereus. *J. Macromol. Sci.*, **B20**, 207 (1981).
61. B. Ivan, T. Kelen and F. Tudos in *Preprints, Eleventh Int. Conf. on Advances in the Stabilization and Controlled Degradation of Polymers*, pp. 61-78, Luzern, May 24-26 (1989).
62. A. Pocker. *Rep. Chem. Sol.*, London, **57**, 165 (1960).
63. R. Schlimper. *Plaste und Kautschuk*, **13**, 196 (1966).
64. K.S. Minsker, Al.Al. Berlin, V.V. Liscitzky, S.V. Kolesov and R.S. Korneva. *Doklady Acad. Nauk SSSR*, **232**, 93 (1977), in Russian.
65. V.N. Myakov. *Doctor Thesis*, Dzerginsk, Kargin's NII Chim. and Technol. Polymers (1978), in Russian.
66. B.B. Troitsky and L.S. Troitskaya. *Uspekhi Khimii*, **54**, 1287 (1985), in Russian.
67. G.A. Razuvaev, L.S. Troitskaya and B.B. Troitsky. *J. Polymer Sci.*, **A-1**, **9**, 2673 (1973).
68. M. Carenza, Yu.V. Moiseev and G. Plama. *J. Appl. Polym. Sci.*, **17**, 2685 (1973).

69. T. Kelen. *IUPAC Makro Mainz, Prepr. Short Commun.*, **1**, p. 5596, 26th Intern. Symp. Makromol., Mainz (1979).
70. B. Ivan, T.T. Nagy, T. Kelen, B. Turcsanyi and F. Tudos. *Polym. Bull.*, **2**, 83 (1980).
71. I. Drerxler and S. Simonik. *Angew. Makromol. Chemie*, **102**, 71 (1982).
72. B.Sh. Bashkirov and A.I. Maklakov. *Highmolec. Comp.*, **A24**, 762 (1982).
73. K.S. Minsker, S.V. Kolesov, V.M. Yanborisov, Al.Al. Berlin and G.E. Zaikov. *Highmolec. Comp.*, **A27**, 1217 (1985).
74. K.S. Minsker and M.I. Abdullin. *Doklady Acad. Nauk SSSR*, **266**, 140 (1982), in Russian.
75. K.S. Minsker, M.I. Abdullin, R.R. Gizatullin and G.E. Zaikov. *Polymer Degradation and Stability*, **11**, 27 (1985).
76. K.S. Minsker, M.I. Abdullin, S.A. Arjakov, V.I. Manushin, A.V. Mamykin and B.A. Kraiikin. *Highmolec. Comp.*, **B21**, 194 (1979).
77. K.S. Minsker, E.I. Kulish and G.E. Zaikov. *Intern. J. Polymeric Mater.*, **24**, 107-110 (1994).
78. A. Jimenez, V. Berenguer, J. Lopez and A. Sanchez. *J. Appl. Polym. Sci.*, **50**, 1565 (1993).
79. A. Jimenez, V. Berenguer, J. Lopez and J. Vilaplana. *J. Appl. Polym. Sci.*, **60**, 2041 (1996).
80. A. Jimenez, J. Lopez, J. Vilaplana and H.J. Dussel. *J. Anal. Appl. Pyrolysis*, **40-41**, 201 (1997).
81. F. Tudos, T. Kelen and T.T. Nagy. *Develop. Polym. Degrad.*, **2**, 187 (1979).
82. A.B. Gagarina, O.T. Kasaikina and N.M. Emanuel. *Doklady Acad. Nauk SSSR, Ser. Chem.*, **249**, 394 (1979), in Russian.
83. O.T. Kasaikina and Le-Ben-Un. *Doklady Acad. Nauk SSSR, Ser. Chem.*, **249**, 394 (1979), in Russian.
84. G. Plama and M. Garenza. *J. Appl. Polym. Sci.*, **16**, 2485 (1972).
85. T.T. Nagy, B. Ivan, B. Turcsanyi, T. Kelen and F. Tudos. *Polym. Bull.*, **3**, 613 (1980).
86. B. Parinet. *J. Chem. Phys.*, **74**, 539 (1977).
87. K.S. Minsker, M.I. Abdullin, V.G. Kalashnikov, V.P. Biriukov, G.M. Trosman and A.V. Salnikov. *Highmolec. Comp.*, **A22**, 213 (1980).
88. B. Ivan, B. Turcsanyi and F. Tudos. *Polymer*, **19**, 351 (1978).
89. T. Kelen, B. Ivan, T.T. Nagy, B. Turcsanyi, F. Tudos and J.P. Kennedy. *Polym. Bull.*, **1**, 79 (1978).
90. K.S. Minsker, S.V. Kolesov and G.E. Zaikov. *Ageing and Stabilization of Polyvinylchloride Based Polymers*, 277p., Nauka, Moscow (1982), in Russian.
91. B. Ivan, T.T. Nagy, T. Kelen, B. Turcsanyi and F. Tudos. *Polym. Bull.*, **2**, 83 (1980).
92. R.P. Wattimer and W.S. Kroenke. *J. Appl. Polym. Sci.*, **25**, 101 (1980).
93. R.P. Wattimer and W.S. Kroenke. *J. Appl. Polym. Sci.*, **27**, 1355 (1982).
94. V. Bellenger. *Europ. Polym. J.*, **28**, 337 (1982).
95. W.H. Starnes and P. Edelson. *Macromolecules*, **12**, 797 (1979).
96. K.Z. Gumargalieva, V.B. Ivanov, G.E. Zaikov, Yu.V. Moiseev and T.V. Pokholok, *Chem. Phys. Reports*, **14**, 965 (1995).
97. B.I. Borisov and N.I. Gromov. *Corrosion and Protection in Oil-Gas Industry*, **3**, 19 (1977).
98. B.I. Borisov. *Colloid J.*, **35**, 140 (1973), in Russian.
99. B.I. Borisov. *Colloid J.*, **3**, 535 (1978), in Russain.

100. E.V. Bukov, L.I. Kromikhfeld, D.D. Rumyantsev and E.A. Fedorovich, *Electrotechn. Industry, Ser. Cable Technics*, **4(182)**, 12 (1970), in Russain.
101. K.S. Minsker in: *Polyvinylchloride and Its Application in Cable Technics*, I.D. Troitsky (Ed.), pp. 152-158, Moscow (1978), in Russian.
102. E.M. Kampouris. *Europ. Polym. J.*, **11**, 705 (1975).
103. E.M. Kampouris. *Polymer Eng. And Sci.*, **16**, 59 (1976).
104. R.S. Barstein, I.N. Afanasieva and L.V. Ryjakoba, *Plastics*, **7**, 73 (1975).
105. N.Yu. Valashik, N.A. Sorokina and R.S. Barstein. *Plastics*, **9**, 32 (1974).
106. E.M. Kampouris. *Polymer Eng. and Sci.*, **16**, 59 (1976).
107. P. Bredereck and H. Endrib. *Plastverarbeiter*, **30**, 77 (1979).
108. B.P. Schtarkmann. *Plastification of Polyvinylchloride*, p. 248, Khimia, Moscow (1975), in Russian.
109. V.K. Milintchuk, E.R. Klinshpont and S.Ya. Pshegetskii. *Macroradicals*, p. 264, Khimia, Moscow (1980), in Russian.
110. V.K. Milintchuk in: *Handbook. Protection from Corrosion, Ageing and Biodeterioration of Machines, Equipement and Buildings*, **2**, pp. 290-312, Machine-Building, Moscow (1987), in Russian.
111. K.S. Minsker, S.V. Kolesov and G.E. Zaikov. *Degradation and Stabilization of Vinylchloride Based Polymers*, pp. 151-153, Pergamon Press, Oxford-New York-Toronto (1988).
112. D. Braun. *Pure Appl. Chem.*, **53**, 549 (1981).
113. R. Renbi and Ya Rabek. *Photodegradation and Photostabilization of Polymers*, p. 234, Mir, Moscow (1978), in Russian.
114. E.D. Owen in: *Develop. Polym. Photochem.*, N.S. Allen (Ed.), **3**, p. 165, Appl. Sci. Publ., London (1982).
115. M.A. Zhuravlev, L.L. Semenytcheva and V.B. Ivanov. *Chemical Physics*, **8**, 790 (1989), in Russian.
116. C. Decker. *Photodegradation of PVC*. In: *Degradation and Stabilization of PVC*, E.D. Owen (Ed.) p. 81, Elsevier, Appl. Sci. Publishers, London-New York (1984).
117. C. Decker and M. Balandier. *Europ. Polym. J.*, **18**, 1085 (1982).
118. N.M. Bitiurin, V.N. Genkin and V.P. Lebedev. *Highmolec. Comp.*, **A25**, 80 (1983), in Russian.
119. N.M. Bitiurin, V.N. Genkin and V.V. Sokolov. *Highmolec. Comp.*, **A24**, 748 (1982), in Russian.
120. Z. Osawa and M. Aiba. *Polym. Photochem.*, **2**, 397 (1982).
121. L.D. Strelkova, G.T. Fedoseeva, V.P. Lebedev, L.A. Tsareva and K.S. Minsker. *Plastics*, **5**, 59 (1983).
122. N.M. Biriutin, V.N. Genkin, V.P. Lebedev, V.V. Sokolov, L.D. Strelkova and G.T. Fedoseeva. *Highmolec. Comp.*, **B24**, 101 (1982), in Russian.
123. C. Decker and M. Balandier. *Polym. Photochem.*, **1**, 221 (1981).
124. J.S. Brooks, R.W. Clarkson, D.W. Allen, M.T.J. Mellor and A.G. Williamson. *Polym. Degrad. Stab.*, **4**, 359 (1982).
125. C. Decker in: *Degradation and Stabilization of PVC*, E.D. Owen (Ed.), p. 314, Elsevier Appl. Sci. Publ., London (1984).
126. E.D. Owen. *ACS Symp. Ser*, **151**, 217 (1981).
127. W.H. Starnes. *ACS Symp. Ser.*, **151**, 197 (1981).

128. B.B. Cooray and G. Scott. *Polym. Degrad. Stab.*, **3**, 127 (1981).
129. G. Scott, M. Taman and J. Vyvoda, *Europ. Polym. J.*, **14**, 1021 (1978).
130. D.Braun and S.Kull. *Angew. Makromol. Chemie*, **85**, 79 (1980).
131. C. Decker and M. Balandier. *J. Photochem.*, **15**, 213 (1981).
132. C. Decker and M. Balandier, *Polym. Photochem.*, **5**, 267 (1984).
133. F. Castillo, F.G. Martiner, R. Sastre, J. Millan, V. Bellenger and J. Verdu. *J. Polym. Degrad. Stab.*, **13**, 211 (1985).
134. V.I. Borisov and V.I. Sushkov. *Highmolec. Comp.*, **A24**, 437 (1982), in Russian.

INDEX